Guidelines for the Identification of Ciliates in Wastewater Treatment Plants

Guidelines for the Identification of Ciliates in Wastewater Treatment Plants

Susana Serrano, Lucía Arregui,
Blanca Perez-Uz, Pilar Calvo and
Almudena Guinea

IWA
Publishing

Published by IWA Publishing, Alliance House, 12 Caxton Street, London SW1H 0QS, UK

Telephone: +44 (0) 20 7654 5500; Fax: +44 (0) 20 7654 5555; Email: publications@iwap.co.uk
Web: www.iwapublishing.com

First published 2008

Cover design by www.designforpublishing.co.uk
Index prepared by Dr Susan Boobis

British Library Cataloguing in Publication Data
A CIP catalogue record for this book is available from the British Library

Library of Congress Cataloging- in-Publication Data
A catalog record for this book is available from the Library of Congress

ISBN: 1843391716
ISBN13: 9781843391715

Contents

Glossary

To help non specialists, a short definition of the most frequently used terms related to ciliates in WWTP is given below.

Adoral zone of membranelles (AZM): Numerous membranelles consecutively arranged from the anterior part of the oral cavity to the left side. Common in heterotrichs and spirotrichs.
Caudal cirri: Cirri placed at the posterior edge of the ventral side of the cell.
Cilium: External motile appendix of eukaryotic cells with a microtubular axoneme; in ciliates, cilia are organised in rows with a coordinated motion.
Cirrus: Specialized cluster of somatic cilia that moves as a unit; typical in sticotrichs and hypotrichs.
Contractile vacuole: Refringent cellular organelle related to osmoregulation. Water is continuously filled and expelled through an expulsion vesicle pore (also called "pulsatile vacuole").
Crawling ciliate: Ciliate that moves on the surfaces with cilia or cirri.

ISBN: 9781843391715. Published by IWA Publishing, London, UK.

Cytophryngeal apparatus (cytopharynx): Cytoskeletal microtubular basket departing from the cytostome into the cytoplasm.
Cytostome: Cell mouth where food vacuoles originate. Located at the cell surface or in the posterior part of the oral cavity.
Extrusome: Membrane-bounded organelle with secretory function.
Fronto-ventral cirri: Cirri placed at the anterior and middle zone of the ventral side of the ciliate cell.
Haplokinety: Row of cilia surrounding the peristome and coming on to the oral cavity of the peritrichous ciliates. It can be considered as a specialized "paroral membrane".
Kinety: Longitudinal or oblique row of somatic cilia.
Larval stage: Motile form of some sessile ciliates.
Lorica: Test, external envelope secreted by some species of different groups of ciliates, sometimes covered by external materials.
Macronucleus: Type of ciliate nucleus involved in the vegetative functions of cells. Its number and morphology are characteristics of each species.
Marginal cirri: Rows of cirri placed at the right and the left margins of the ventral side of the ciliate cell.
Membranelles: Components of the oral ciliature of some groups of ciliates, composed by two to four short rows of cilia (commonly three), transversally placed at the left side of the oral cavity.
Micronucleus: Type of ciliate nucleus holding the complete genetic information of the species, it generates macro and micronuclei during the conjugation.
Myoneme: Contractile microfibrillar bundles in heterotrichs (body contractility) and peritrichs (zooid contractility).
Nassa: Complex cytopharyngeal apparatus in phyllopharyngids and nassophorids, constituted by microtubular bundles.
Nuclear dualism: Two different types of nuclei (macro and micronucleus) with different functions in the cell.
Opercule: Cap or flap covering the lorica or the cyst. Also, cytoplasmic peristomial extension in peritrichs ciliates.
Oral cavity: Cortical funnel that finishes in the cytostome present in some groups of ciliates.
Oral ciliature: Rows or groups of cilia placed around the cytostome or in the oral cavity. Usually related with feeding.
Oral overture: Superficial area of the oral cavity.
Paroral membrane: Row/s of cilia lying at the right edge of the oral cavity in some groups of ciliates. Also named "undulating membrane".

Peniculus: Component of the oral ciliature constituted by three or four longitudinal rows of cilia placed at the left wall of the oral cavity in some groups of oligohimenophorids.

Peristomial lip: External peristomial swelling existing in some peritrichous ciliates.

Polykinety: Oral structure composed by three ciliary rows surrounding the peristome of the peritrichs and originating the first peniculus in the oral cavity. Located behind the haplokinety.

Preoral kinety: Oblique kinety with longer cilia close to the cytostome of some phyllopharingids as *Chilodonella*.

Saprobic system: Group of species that characterizes environments with different organic pollution.

Sessile ciliate: Cell attached by stalks, loricas, etc. to surfaces or living organisms.

Somatic ciliature: Rows or specialized groups of cilia placed on ciliate surfaces. Commonly related with cell motility.

Spasmoneme: Contractile microfibrillar bundle in the stalk of some peritrichs.

Stalk: Threadlike supporting structure emerging from the posterior part of the cell and related to the adhesion to surfaces, for example peduncule of peritrich and suctorid ciliates.

Swimming ciliate: Motile ciliate that moves freely in the water.

Toxicyst: Tubular or ovoid toxic extrusome typically present in predatory ciliates.

Transversal cirri: Cirri placed at the posterior zone of the ventral side of the ciliate cell.

Trichocyst: Subpellicular extrusome spindle-shaped found in several groups of ciliates.

Vestibulum: Oral depression in some ciliates.

About the Authors

The authors belong to a research group working on protists at the Universidad Complutense de Madrid. This group was founded by Prof. Dimas Fernández-Galiano about thirty years ago. Traditionally, the research carried out by the group has been based on taxonomy, morphogenesis and cytoskeletal components of ciliates. Taking into account our knowledge about the cell structures and the taxonomy of ciliates, a line of research was set up in 1990, to study the structure and function of protists in wastewater treatment plants, their relationships to physical-chemical and control parameters in these artificial ecosystems, and the role of these micro-organisms as bio-indicators in wastewater treatment plants. Numerous publications about ciliates, their taxonomy and the importance in Wastewater Treatment Plants, have been produced at the specialized protistological and water journals.

Acknowledgements

This work was supported by grants (INV.PR.0.0707.15396) of the spanish Universidad Complutense de Madrid and (BOS2002-01042) Ministerio de Ciencia y Tecnología.

Microphotographs Collaborators

We would like to thank A. Zornoza (AVSA-EGEVASA, WWTP Quart-Benager, EPSAR); N, Fernández, L. Isac, E. Rodriguez (GBS); P. Fontela, S. Ruiz (Universidad de Santiago de Compostela); I. Campos (Universidad Complutense de Madrid) and W. Foissner (Universität Salzburg) for generously providing several ciliate species photographs.

1

Introduction

1.1 WASTEWATER TREATMENT PLANTS: BIOLOGICAL DEPURATION

Water contamination and problems derived from it are amongst the main priorities of environmental administrations in the world today. Different systems of wastewater management have been developed in the last decades, most of them including biological activities as efficient tools for the removal of organic material and other chemical and biological contaminants.

The biological reactors in Wastewater Treatment Plants (WWTP) are artificial ecosystems projected to achieve the development of bioaggregates (flocs and biofilms) in which stable communities of organisms, mainly bacteria and protozoa, are responsible for the removal of organic and inorganic pollutants (Bick 1963; Bishop and Kinner 1986, Madoni 1988, 1994b; Bitton 2005).

Microbiota of activated sludge systems, mainly flocculating bacteria and protozoa, has been extensively studied (Curds and Cockburn 1970a; Madoni *et al.* 1993; Madoni 1994a; Martín-Cereceda *et al.* 1996). Filamentous bacteria,

© 2008 IWA Publishing. *Guidelines for the Identification of Ciliates in Wastewater Treatment Plants* by Susana Serrano, Lucia Arregui, Blanca Perez-Uz, Pilar Calvo and Almudena Guinea.
ISBN: 9781843391715. Published by IWA Publishing, London, UK.

protozoa and metazoa are the main communities described in biofilms in rotating biological contactors (Pretorius 1971; Torpey *et al.* 1971; Madoni *et al.* 1979; Hoag *et al.* 1980; Madoni and Ghetti 1981; Kinner and Curds 1987; Kinner *et al.* 1983, 1988; Chung and Strong 1991; Madoni 1994a). Floc and biofilm formation facilitate the separation of bioaggregates from purified water by deposition or adherence to solid surfaces, making possible the clarification of the effluent. Settled sludge is also used in activated sludge processes to reinoculate biological reactors.

1.2 CILIATES IN WWTP BIOLOGICAL REACTORS

Ciliates are a highly diverse group of protozoa very common in biological reactor communities (Curds and Cockburn 1970a; Madoni 1991). Excellent manuals of protozoa and ciliates are available (Curds 1975; Puytorac *et al.* 1987; Sleigh 1989; Lynn and Corliss 1991; Haussman and Bradbury 1996; Lynn and Small 2000; Hausmann *et al.* 2003). The main characteristics of this group of microorganisms can be summarized as follows:

- cells present cilia; ciliature can be absent in vegetative cells of some groups but in those cases it can be observed in larval stages. Somatic ciliature is organized in longitudinal or oblique ciliary rows (kineties) and is related to the cellular motion. Oral ciliature is found around the cytostome or within the oral cavity, giving rise to a flow of water and particulate food towards the cytostome.
- phagotrophic cells with a cytostome (rarely without one), associated with specialized cytoskeletal cytopharingeal structures. Digestive vacuoles are usually found in the cytoplasm.
- two types of nuclei (nuclear dualism): macronucleus (large in size, with different shapes; one or more per cell) and micronucleus (small in size, spherical or ovoid in shape; one or more per cell). Macronucleus is involved in the control of vegetative functions and micronucleus represents the storage of the genetic information, and it is the active nucleus during the conjugation process.
- they contain one or more contractile vacuoles managing osmotic regulation processes with cortical water-expulsion pore (frequently with a precise location within the cell).
- asexual reproduction by transversal binary fission as the most common growth mechanism. Sexual phenomenon (conjugation) is also present in some species.
- terrestrial and aquatic organisms (freshwater, seawater and wastewaters) taking part of biological food chains (Fenchel 1987).

Ciliates have an essential contribution in depuration processes because they are involved in the transfer of organic and inorganic compounds through trophic webs (Dive 1973; Small 1973; Anderson 1988; Cairns and Pratt 1988; Curds 1992; Madoni 1994b; Sommer 1996; Mas-Acebes 2007) and in the elimination of organic matter either dissolved (Güde 1979) or flocculated (Ratsak *et al.* 1994). Moreover, the predatory activity of bacterivorous ciliates maintains the balance of bacterial populations (Curds and Fey 1969; Curds 1975; Mas-Acebes 2007) contributing also to the removal of bacterial pathogens (Curds and Fey 1969). Additionally, bioaggregation due to ciliate activities has been described (Watson 1945; Curds 1963; Arregui *et al.* 2007). Therefore the presence of these organisms plays a role in effluent clarification (Curds and Vandyke 1966; Curds *et al.* 1968; Curds 1975; Ratsak *et al.* 1994).

Many WWTP studies have been done about ciliate diversity in different activated sludge systems (Madoni 1982; Esteban *et al.* 1990, 1991; Martín-Cereceda *et al.* 1996) and rotatory biological contactors (Cybis and Horan 1997; Pérez-Uz *et al.* 1998; Martín-Cereceda *et al.* 2001a, b; Fried and Lemmer 2003).

Species reported inside the biological reactors can be divided into two groups according to their ecological niche (Poole 1984; Madoni 1988; Martín-Cereceda *et al.* 1996, 2001c, d):

i. related to the flocs or biofilms (autochthonous communities); or
ii. related to the mixed liquor (transitional populations).

Autochthonous ciliates are adapted to live attached to the floc/biofilm or close to its surface and are named as sessile or crawling ciliates respectively. Sessile ciliates are fixed through stalks or adhesive structures, whilst crawling ciliates present flattened bodies and a specialized ventral ciliature allowing movement on the surface. Pedunculates feed on suspended bacteria whilst crawling ciliates mainly feed on flocculating bacteria. Higher abundances correspond to these categories.

Transitional ciliates are free-swimming in mixed liquor. Non-associated to the floc/biofilm ciliates, they also contribute to the elimination of suspended bacteria (including pathogens). Higher abundances of these ciliates are associated with start up or overloading of the WWTP processes.

1.2.1 Ciliates as Bioindicators in WWTP

Presence and abundance of ciliates in WWTP depends on the physical-chemical conditions, so changes in the environmental circumstances induce changes of the community structure. Many experimental studies have demonstrated the indicator value of ciliates in different habitats (Bick 1963; Cairns 1978; Al-Shahwani and Horan 1991).

Structure, abundance and composition of the groups or ciliate species in flocs or biofilms can supply information about the efficiency of the depuration process. Ciliates are reported by various authors as bioindicators of the performance of biological reactors currently being used by lab technicians in the routine microscopical controls in WWTP (Curds 1969; Curds and Cockburn 1970b; Foissner 1988; Madoni 1982, 1988, 1994a, c; Foissner and Berger 1996; Martín-Cereceda *et al.* 1996). Accuracy in the species characterization/ identification and diagnosis of the groups of ciliates that can be used as biological indicators of the depuration process is indispensable (Madoni 1988, 1991; Salvadó 2001). Madoni (1994a, c) established a biological index to evaluate the biological functioning in activated sludge plants – the "sludge biotic index (SBI)"- being based on microscopic identification and estimation of protozoa abundance (small and large flagellates, naked and testaceae amoebas and groups and species of ciliates). This index is used by almost all WWTP controllers.

1.3 CLASSIFICATION OF CILIATES

Most ciliate taxonomic studies are directed at specialists, and few authors have proposed keys for the identification of ciliates in wastewater treatment plants. Several specific works to be mentioned are:

- Curds (1969) *An illustrated key to the British freshwater ciliated protozoa commonly found in activated sludge*. This key is based on the previous (valid at that time) proposal of the Committee on Taxonomy and Taxonomic Problems of the Society of Protozoologist (Honigberg *et al.* 1964) in which ciliates appearing in wastewaters were included in four subclasses: Suctoria (sessile ciliates without cytostome or ciliature at the adult stages, with tentacles), Peritrichia (stalked ciliates without somatic ciliature at the adult forms but with oral ciliature), Holotrichia (swimming ciliates with homogeneous somatic ciliature and different patterns of oral ciliature) and Spirotrichia (heterotrichs and the "old" crawling hypotrichs with somatic ciliature and adoral zone of membranelles at the left of the oral cavity). Curds later published a more general book about the characterization of freshwater genera of ciliates (Curds 1982).
- Madoni (1988) published a useful manual, *I protozoi ciliate nel controllo di efficienza dei fanghi attivi*, in which practical methodological information is provided as well as a simple key for the classification of activated sludge plants ciliates. This guide used the Committee on Systematics Evolution of the Society of Protozoologists classification (Levine *et al.* 1980), considering three classes: Kinetofragminophora (simple oral ciliature around the cytostome), Oligohymenophora (with

oral ciliature inside an oral cavity and constituted by few elements) and Polyhymenophora (with oral ciliature inside an oral cavity constituted by numerous elements). Keys for representative genera and short descriptions of species were also included.

Other works have included morphological descriptions for identification of ciliate species in WWTP (Curds 1975, 1982, 1983; Eikelboon and Buijsen 1981; Berger *et al.* 1984; Borror and Hill 1995; Fernández-Galiano *et al.* 1996).

Recently, taxonomic classification of ciliates has changed according to the new structural, morphogenetic, molecular and phylogenetic data (Margulis *et al.* 1990; Sogin 1991; Fleury *et al.* 1992; Schlegel 1994; Lynn 1996; Corliss 1998; Lipscomb *et al.* 1998; Bernhard *et al.* 2001; Cavalier Smith 1998, 2002). According to phylogenetic systematics, the new classification considers that complex oral structures are a characteristic of primitive/earlier groups (plesiomorphies), whilst simplified oral structures are derived (apomorphies) in the evolutionary process.

The Last edited proposal of the Society of Protozoologists (Lee *et al.* 2000) takes into account ultrastructural and phylogenetic studies as well as other taxonomical characteristics such as morphology, morphogenesis, ecology, etc. The Phylum is divided in two Subphylum and ten Classes:

- Subphylum Postciliodesmatophora: somatic dikinetids with overlapping postciliary microtubular ribbons also constituting the cytopharingeal apparatus, presence of ZAM in some groups, macronuclear division directed by microtubules outside the macronuclear membrane; includes kariorelictids (e.g. *Loxodes*) and heterotrichs (e.g. *Spirostomum* or *Stentor*).
- Subphylum Intramacronucleata: macronuclear division is controlled by intramacronuclear microtubules; includes diverse groups such as spirotrichs (e.g. *Halteria*, *Oxytricha* or *Aspidisca*), Litostomatids (e.g. *Enchelys*, *Litonotus*), Phylopharyngids (e.g. *Chilodonella* or suctorids), Nassophorids (e.g. *Drepanomonas*), Colpodids (e.g. *Colpoda*), Prostomids (*Coleps* or *Prorodon*), Plagiopylids (e.g. *Plagiopyla*), and Oligohimenophorids (e.g. *Tetrahymena*, *Paramecium*, *Uronema*, peritrichous ciliates).

Other taxonomic works (Haussman *et al.* 2003) and the recent "revolutionary" classification of the eukaryotic organisms with an emphasis on protists - proposed by Adl *et al.* 2005 – have also been taken into account in this work. The last revision (Adl *et al.* 2005) underlined recent phylogenetic and ultrastructural criteria, proposing six clusters of monophyletic groups of living organisms – multicellular eukaryotes and protists are grouped together – from which they had emerged. Ciliates are included within the Chromoalveolata with parasitic apicomplexes, dinoflagellates and other microorganisms that present alveoli under the cellular membrane.

The classification in this book is based on that proposed by the Society of Protozoologist (Lee *et al.* 2000) with Postciliodesmatophora: Kariorelictids and Heterotrichs and Intramacronucleata: Spirotrichs, Armophorids, Listostomatids, Phyllopharyngids, Nasophorids, Colpodids, Prostomids, Plagiopylids and Oligohimenophorids (including peniculids, scuticociliates, hymenostomatids, apostomatids, astomatids, and peritrichous ciliates).

The excellent work on ciliates by Foissner *et al.* (1991-1995), in which a detailed description of a great variety of species of saprobic systems is given, should also be highlighted.

The main purpose of this work is to provide an up-to-date and easy to use key including simple morphological characteristics that can be used by technicians and non-specialists in ciliate taxonomy in order to accurately identify taxonomic groups and species of ciliates typical in WWTP and other aquatic environments.

2

Methodologies to study ciliates in Wastewater Treatment Plants

2.1 INTRODUCTION

The main methodologies used to obtain reliable results for the use of ciliates as tools in wastewater technologies involve all the mechanisms to study the community structure including sampling, counting and identifying. It can not been stressed enough that, for this purpose, deficiencies in any of these steps will clearly limit the usefulness of these microorganisms as predictive tools for any ecosystem studied.

This chapter focuses on a brief description of the most common methods for sampling, counting and identifying ciliate populations from the biological process in a WWTP. Methodologies used are common to those employed in other aquatic ecosystems, but here must be adapted to the particular characteristics of these environments, such as high organic load, high density of bacterial populations as well as a diverse and abundant ciliate community generally present in a healthy

ISBN: 9781843391715. Published by IWA Publishing, London, UK.

aerobic biological process. Finally, a brief account of published methodologies to assess the condition of a wastewater treatment process with ciliates will be shown.

2.2 SAMPLING CILIATE POPULATIONS IN WWTP

The strategy to sample ciliates from WWTP is focussed on the recovery of the whole diversity of these microorganisms present in the ecosystem to obtain quantitative and reliable taxonomic data required in ecological studies.

Protozoan abundance in the biological treatment is normally over 10^4 cells/ml, and ranges from 10^5–10^6 cells/ml depending on the type of process and its state (Madoni 1988). In consequence, sample volume does not need to be large and 25–100 ml would be enough to carry out the study of the ciliate populations. To preserve the populations in a similar state to the natural conditions, containers should be large enough to keep an air chamber on top of the sample (at least 1/3 of the container volume) and samples should be kept cool and away from direct sunlight until processing in the laboratory. Long transportation or storage times could contribute to changes in both physical–chemical (e.g. temperature, pH, oxygen) and biological (e.g. encystment–excystment, grazing) characteristics and therefore to a loss of sensitive species and an unreal quantification of ciliates.

Wastewater samples are normally not fixed since flocculent material would precipitate preservatives and would also severely interfere with the observations, identification and counting of protozoa. This is due to several factors also applicable to natural samples: smaller protozoa may be hidden by floc aggregates, fixed ciliates are very difficult to identify accurately, and finally many fragile ciliates will burst with the most commonly used preservatives. Wastewater samples are thus kept in living conditions, and they should be processed as soon as possible and within twelve hours of sampling. Oxygen depletion can be avoided, once in the lab, using mechanical agitation of the sample with a magnetic stirrer, a shaker or an aquarium air pump; this will avoid death of those most oxygen sensitive ciliates.

2.2.1 Sampling activated sludge WWTP

The microbial community of activated sludge plants develops on suspended flocs or freely in the mixed liquor. These flocs have abiotic – both organic and inorganic – and biotic components comprising bacteria, fungi and other organisms such as protozoa or small invertebrates. To sample all groups of ciliates in these artificial ecosystems, both flocs and mixed liquor are sampled at the same time submerging a flask in the aeration tank where the biological treatment takes place. Once the sample is taken, it should be kept with mechanical agitation or aeration using an aquarium air pump to keep flocs aerated and to avoid settling of suspended flocs.

2.2.2 Sampling fixed biofilm WWTP

The microbial community of biofilms develops on an inert surface. Protozoa are an integral part of this biofilm and grow directly as attached or crawling forms integrated intimately within the biofilm (Kinner *et al.* 1983; Kinner and Curds 1987; Martín-Cereceda *et al.* 2001c). Therefore, in these cases the sampling unit must recover a defined volume of biofilm to retrieve the microorganisms within it. Scraping a complete section of the biofilm from the inert surface that support it or detaching the biofilm manually (or automatically) by shaking the support are the most common described methodologies in the literature (Selivanovskaya *et al.* 1997; Pérez-Uz *et al.* 1998; Martín-Cereceda *et al.* 2001a, b, c; Vymazal *et al.* 2001; Puigagut *et al.* 2007). The best way to determine the biofilm accurately would be to mark a defined area of the biofilm to be removed. The biofilm is then detached and dispersed in a defined volume of filtered wastewater (e.g. 0.45 μm filtered) from the same sampling point. To quantify populations it is important to use a calibrated container so that biofilm weight (or volume) is known. The assessment of biofilm volume can be recorded in a calibrated measuring cylinder registering dilution volume changes (Kinner and Curds 1987; Kinner *et al.* 1988). The dilution volume of wastewater for the sample will be related to biofilm weight and defined for the purpose specifically (e.g. 1:4, 1:6, 1:10, etc). Samples must be kept with mechanical agitation once in the lab.

Replicate samples of biofilm would also be necessary to assess dry-weight (Selivanovskaya *et al.* 1997) or total solids (TS) and volatile solids (VS) if ciliate abundance or biomass is going to be related to these biofilm parameters (Pérez-Uz *et al.* 1998; Vymazal *et al.* 2001; Martín-Cereceda *et al.* 2001a, b, c; Martín-Cereceda *et al.* 2002).

2.3 COUNTING CILIATES IN WWTP

Enumeration of ciliates in WWTP is a mandatory procedure to evaluate these microorganisms as indicators of effluent quality or as a biological process management control.

The methodology is direct counting that has to be done as soon as possible after sample collection to minimize changes during processing. Enumeration of living cells is done by direct microscopy on aliquots of unfixed samples. It is advisable to prepare an inventory of ciliate species present before counting. Since ciliate numbers are usually not very large in the biological process, counting chambers are generally not necessary, and small aliquots taken with a calibrated pipette are mounted and counted on a slide under a coverslip (Figures 2.1–2.5).

Counting has to be carried out relatively quickly, since drying out of the sub-sample will involve eventual bursting of those more delicate cells and will therefore lead to errors in the counting procedure; this is one of the main reasons to use small coverslips. When the number of ciliates is quite large and counting time is foreseen to take longer, a sort of chamber could be prepared sealing the borders of the coverslip with a fine line of vaseline before mounting it on the calibrated sub-sample volume (Figure 2.5); this would avoid drying out of the sample although the heating of the sample, through longer observation times under the microscope, can not be avoided unless a cold stage is used.

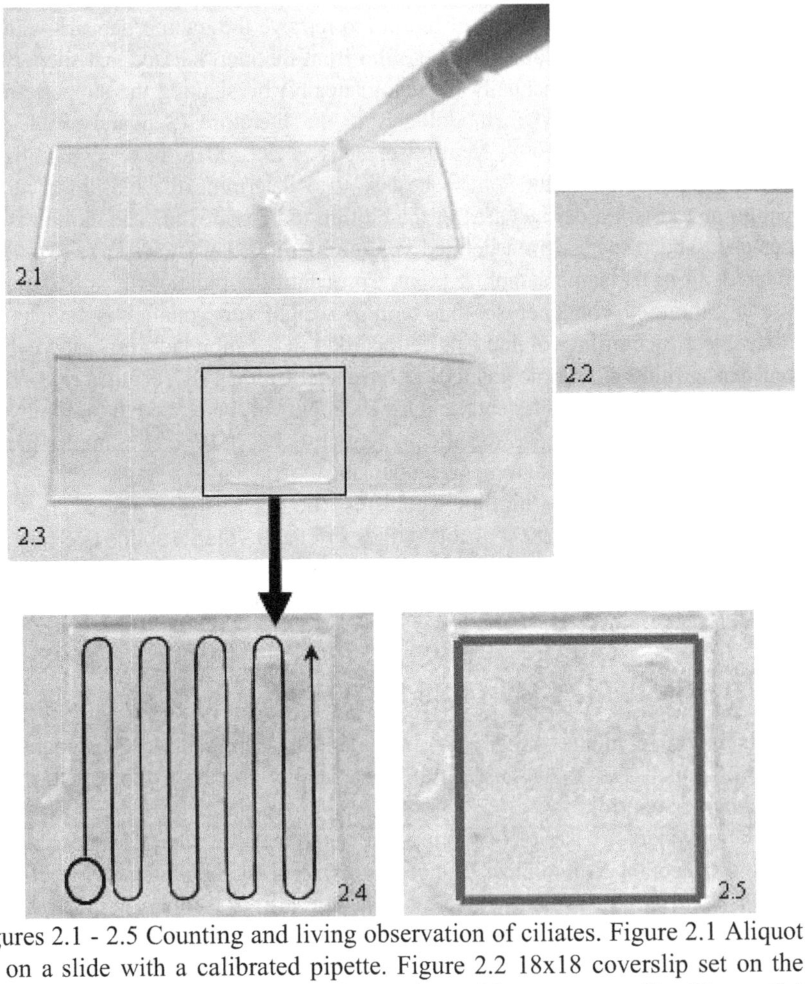

Figures 2.1 - 2.5 Counting and living observation of ciliates. Figure 2.1 Aliquot set on a slide with a calibrated pipette. Figure 2.2 18x18 coverslip set on the sample aliquot. Figure 2.3 First observation without a coverslip. Figure 2.4 Counting plan to cover the complete 18x18 sample area. Figure 2.5 Grey lines show the position of vaseline lining between slide and coverslip to avoid sample evaporation.

Table 2.1. Sub-sample counting methodology used by previous authors in wastewater studies on activated-sludge (AS), rotating biological contactors (RBC).

Reference	WWTP	Sub-sample Counting methodology		Other details
		Volume (µl)	Replicates	
Curds *et al.* 1968	AS/TF	≈28	18	Slide/coverslip
Madoni & Antonietti 1984	AS	250	4	Slide/coverslip
Madoni 1988	AS	25	2-4	Slide/coverslip
Esteban *et al.* 1990	AS	50	5	Slide/coverslip
Augustin & Foissner 1992b	AS	10	5	Thoma chamber
Madoni *et al.* 1993	AS	25	2	Slide/coverslip
Salvadó 1994	AS	50	4	Slide/coverslip
Fernández-Galiano *et al.* 1996	AS	25	3	Slide/coverslip
Kinner & Curds 1987	RBC	20	3	Dotted Slide/coverslip
Kinner *et al.* 1988	RBC	40	3	Slide/coverslip
Luna-Pabello *et al.* 1996	RBC	100	1	Slide/coverslip
Pérez-Uz *et al.* 1998	RBC	25	3	Slide/coverslip

In general, most studies follow a protocol to count protozoa within 30 minutes to 12 hours from sampling (Augustin and Foissner 1992b; Fernández-Galiano *et al.* 1996; Martín-Cereceda *et al.* 2002; Lee *et al.* 2004). The counting area is defined by 18x18 mm coverslips and counting should be done with 100x or 200x microscope magnifications. Differences in the literature can be found, however, in the sub-sample volume to be counted (Table 2.1). The selected volume and the number of replicates depends on the number of cells present, since a minimum number of cells must be counted according to the Poisson model distribution so that the experimental mean can be considered statistically precise and within confidence limits.

Madoni (1984) described a sub-sampling technique and determined that aliquot volume must contain at least 20 individuals for each species to keep the variation coefficient and error low; in that study, for example, four replicates of 25µl or one of 50µl would be enough for most representative species. Obviously, larger volumes with higher cell abundances will involve less replicates that those with lower volumes and with fewer cell abundances (Table 2.1). In any case, decisions on volume and replicates are a balance between degree of precision and expenditure of time (Madoni 1984; Madoni 1988; Augustin and Foissner 1992b) and must be assessed for the specific study and the type of plant (Madoni 1984).

2.4 IDENTIFICATION OF CILIATES

Ciliate identification requires both live and stained observations. Live observation is important because some characteristics are only observed in this state; staining is necessary to identify ciliates to species level. Staining shows the arrangement of the ciliature and infraciliature in somatic and oral structures, and these characteristics are almost always mandatory to identify the species.

Different techniques can be used to stain ciliated structures. The classic techniques have involved the use of different silver salts that generally precipitate specifically on microtubular structures, although sometimes other type of structures can be observed. Recently, techniques based on hibridization of molecular probes against microtubular structures have been successfully developed; however, the lack of comparative taxonomic studies on ciliates with these have still limited their use in identification.

2.4.1 Living observation

Live cell observation is important in the identification of ciliates sometimes even to genus level, since characteristics such as shape, movement, colour or certain structures (nuclei, oral area, contractile vacuoles, ciliature, etc) are only visible when the organism is alive. To observe these characteristics *in vivo* it is necessary to use both bright field and contrast microscopy (phase contrast or differential interference contrast –Nomarski). Observation under the microscope should be carried out first in a small aliquot on a slide without a coverslip (Figure 2.2) since large ciliates are deformed when compressed under a coverslip. Next, a coverslip can be set as a chamber with vaseline (petroleum jelly) lining the borders or directly on the aliquot (Figure 2.3), and then more detailed observations on nuclei or surface characteristics such as granules, striations, trichocysts or ciliature can be observed in detail (Figures 2.6–2.13). Sometimes, it is advisable to use a very small sample quantity to observe these small details, or to leave it to evaporate so that the coverslip traps the ciliate underneath. At this stage, the ciliate will obviously be deformed but some structures will be more easily observed in these conditions.

Figures 2.6 – 2.13. (next page). Living observation: shape and morphology with phase and interference contrast microscopy. Figure 2.6 *Tokophrya quatripartita*. Figure 2.7 *Litonotus lamella* oral region. Figure 2.8 *Uronema nigricans* oral and somatic ciliature. Figure 2.9 *Euplotes patella* ventral somatic ciliature. Figure 2.10 *Vorticella fromenteli* cortical morphology (striation and cortical granules). Figure 2.11 *Epistylis chrysemidis* peristomial area (kindly provided by A. Zornoza). Figure 2.12 *Colpidium colpoda* general view and oral area. Figure 2.13 *Trithigmostoma srameki* internal structures.

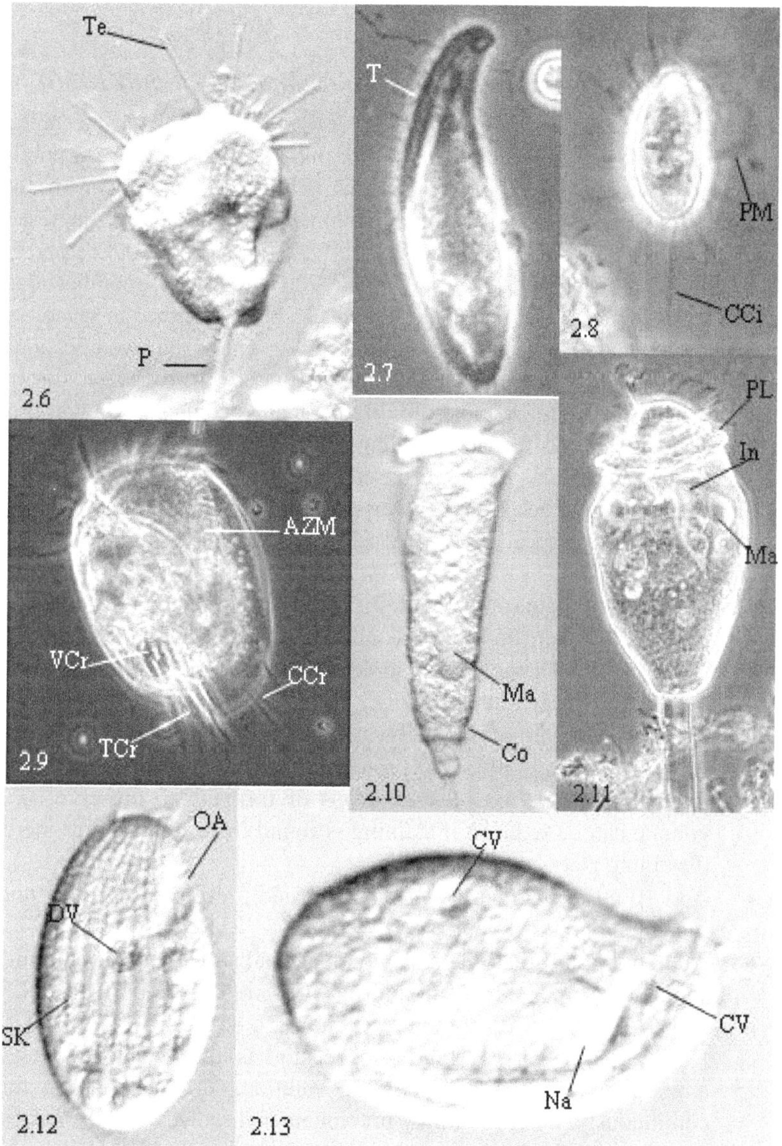

(AZM) adoral zone of membranelles; (CCi) caudal cilium; (CCr) caudal cirri; (Co) cortical morphology; (CV) contractile vacuole; (DV) digestive vacuole; (In) infundibulum; (Ma) macronuclei; (Na) nassa; (OA) oral area; (P) peduncule; (PL) peristomial lip; (PM) paroral membrane; (SK) somatic kinety; (T) trichocysts; (TCr) transverse cirri; (Te) tentacles; (VCr) ventral cirri.

2.4.2 Silver staining techniques

2.4.2.1 Silver carbonate staining (Fernández-Galiano 1966)

Fernández-Galiano (1966, 1976) described the use of this histological technique on ciliates for the first time. Since then, several modifications have been published, concerning: staining these microorganisms from specific ecosystems, those showing difficulties to be stained with the original technique, and for single cells or for small sample volumes (Augustin *et al.* 1984; Foissner 1991; Lee and Soldo 1992; Olmo and Téllez 1997; Ma *et al.* 2003). Specific modifications for wastewater treatment ciliate populations have also been described in several works (Fernández-Galiano 1994; Esteban *et al.* 1998). The main modifications for wastewater ciliates rely on the elimination of excess organic matter that might interfere with a successful staining of ciliates. Organic compounds might produce precipitation of silver carbonate avoiding the deposition of these salts on the microtubular structures and, therefore, no staining or very faint staining is observed.

Impregnation can be used for large or small volumes. For larger volumes, the technique, as we use it specifically for wastewater ciliate populations, is briefly as follows:

- Fix 5 ml of ciliate sample with 2–3 drops of commercial formalin (37%) in a small beaker (Figure 2.14); allow fixation for 2 to 3 minutes.
- Wash the fixed sample by mild centrifugation and resuspension of cells with distilled water at least twice.
- Resuspend in 5 ml of distilled water and again add 2 drops of formalin.
- Add 22–25 drops of a 4% proteose peptone solution (100 ml of 4% Bacto Peptone Difco plus 25 drops of formalin to preserve it). This volume can be reduced if staining is found to be too faint, or increased if staining is too strong.
- Add 10 ml of distilled water and then add 5 drops of pyridine under a fume hood, mixing well.
- Add 2 ml of a Rio Hortega´s ammoniacal silver carbonate solution. This solution is prepared by mixing a silver nitrate solution (50 ml, 10% $AgNO_3$) with a sodium carbonate solution (150 ml, 5% $Na_2CO_3.10H_2O$); this cloudy suspension is then clarified by adding ammonium hydroxide (NH_4OH) solution, drop by drop, mixing continuously. Once the milky precipitate is dissolved and the solution is transparent add sodium carbonate solution (5% $Na_2CO_3.10H_2O$) to reach 750 ml. This solution must be stored in dark containers.
- Heat the mixture in a water bath at 65°C until a change in the colour from transparent or milky to a transparent dark-brown or even black.

- Stop the staining by cooling straight away by pouring the mixture into 10-20 ml of cold distilled water in a larger evaporating dish. The original version indicates stopping the reaction with a 5% sodium thyosulphate ($Na_2S_2O_3$) solution; however, if the observation of ciliates is not going to be made straight away, it is better to stop the reaction with just cold water, and in this way the staining is kept in a good observable condition for longer. Leave cells to settle and collect them with a micropippete for observation.

Alternatively, lower sample volumes or single cells can be processed in depression slides or in watch glasses (see Figure 2.14. d, e). Sample fixation and washing can be carried out manually with a micropipette. Reagents, in this case, can be mixed in a small 25 ml beaker: 1 ml of distilled water, 2 drops of formalin, 22-25 drops of protease-peptone solution, 5 drops of pyridine, 10 ml distilled water and 2 ml of ammoniacal silver carbonate solution. After mixing well, the mixture is added in excess (e.g. 1 ml in a depression slide) to the fixed-washed ciliates previously concentrated in a small drop in the watch glass. Heating is also carried out in a water bath at 65ºC.

This technique is simple, easy and fast to carry out. Permanent slides can be prepared, dehydrating and mounting the stained ciliates; however, the quality of observations will be worse than those made on freshly stained cells. The impregnation stains infraciliary structures, such as kinetosomes and associated fibers (kinetodesmal fibers), and it may also stain parasomal sacs, myonemic systems and spasmonemes as well as nuclei, vacuoles, contractile vacuole pores, etc (Figure 2.4.a, b, c).

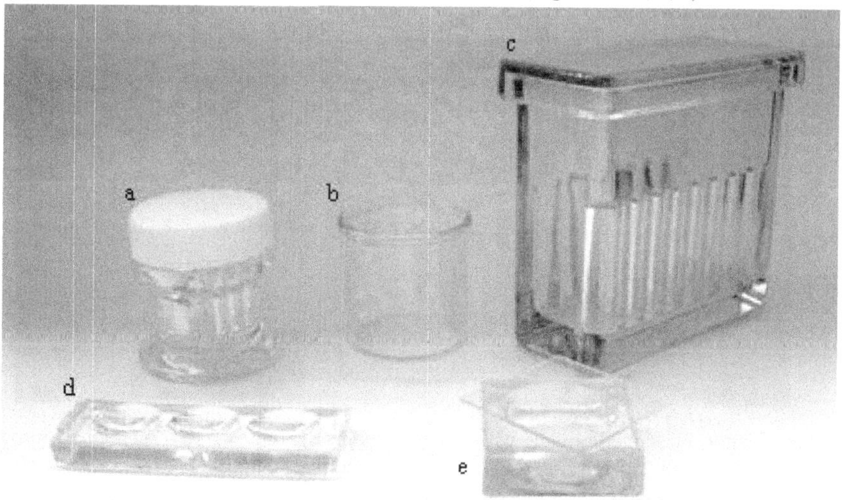

Figure 2.14. Glassware for staining procedures. (a) Columbia jar to process coverslips, (b) Small 25 ml beaker, (c) Hellendahl staining jar to process slides, (d) 3-wells depression slide (1 ml), (e) Watch glass.

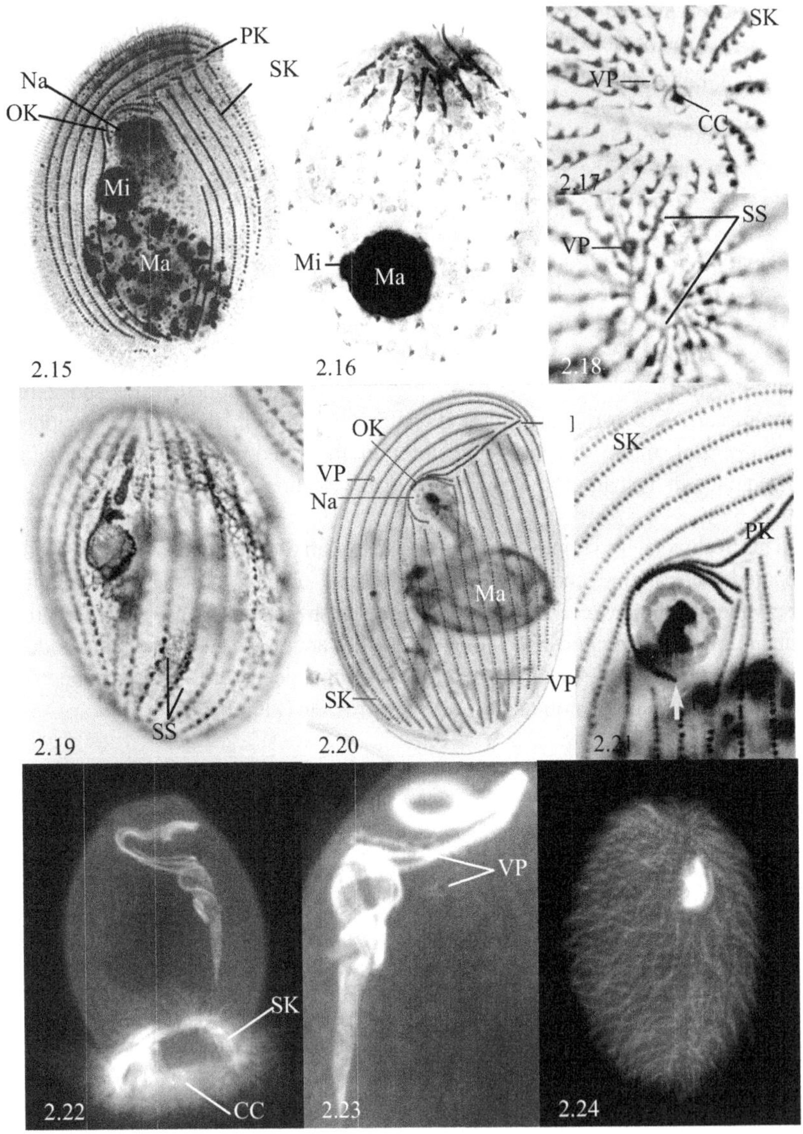

Identification of stained ciliates with silver carbonate (Figures 2.15-2.17), silver nitrate (Figures 2.18 and 2.19), protargol (Figures 2.20 and 2.21) and Flutax (2.22–2.24). (**Full captions on following page**). Key: (CC) caudal complex; (Ma) macronucleus; (Mi) micronucleus; (Na) nassa; (OK) oral kinety; (PK) preoral kinety; (SK) somatic kinety; (SS) silverline system or argyrome; (VP) vacuole pore.

(Figures 2.15-2.24 on previous page). Figure 2.15 *Pseudochilodonopsis similis* ventral infraciliature. Figure 2.16 *Coleps* sp. somatic and oral infraciliature. Figure 2.17 *Urocryptum tortum*, detail of the caudal area. Figure 2.18 *Urocryptum tortum*, detail of caudal area, silverline system. Figure 2.19 Ventral oral view of *Urocryptum tortum*. Figure 2.20 *Trithigmostoma cucullulus* ventral somatic and oral infraciliature (kindly provided by Prof. W. Foissner). Figure 2.21 *Trithigmostoma cucullulus*, detail of oral ciliature (kindly provided by Prof. W. Foissner). Figure 2.22 Telotroch larval stage of *Opercularia coarctata* oral infraciliature in the closed infundibulum and somatic infraciliature. Figure 2.23 Detail of the oral infraciliature. Figure 2.24 *Tetrahymena thermophyla* somatic infraciliature.

2.4.2.2 Silver nitrate staining (Chatton and Lwoff 1930)

Silver nitrate staining is a technique originally described by Chatton and Lwoff (1930, 1936) without any previous desiccation of cells, which is why this technique is referred to as "wet" silver nitrate staining. The technique has been modified or slightly adapted several times (Corliss 1953; Roberts and Causton 1988; Foissner 1991). We generally use the modification by Roberts and Causton (1988), which is processed on coverslips rather than on slides and is briefly as follows:

- Concentrate a ciliate sample of 12 ml approximately by centrifugation and then fix cells under a fume hood with 1-2 ml of Champy's fluid (7 parts 1% CrO_3 : 7 parts 3% $K_2Cr_2O_7$: 4 parts 2% OsO_4; add OsO_4 just before use) mixing well and allowing fixation for 3 minutes.
- Resuspend with Da Fano's fixative (1% $CoNO_3$, 1% NaCl, 10% formalin) and leave for 10 minutes. Wash twice with tap water, leaving at least 1 minute each time.
- Mount cells in a 15% gelatine–Da Fano's mixture (2:1), mixing the concentrated cells (in less than 0.5 ml) with 4–5 drops of melted gelatine-Da Fano´s mixture (keep in a water bath). This process has to be carried out fast since the mixture solidifies very quickly when it cools. To avoid hardening, dispense a very small drop of the cell suspension-gelatine–Da Fano´s mix on one side of a warm coverslip, allow a mounted needle parallel to the coverslip to contact the drop of gelatine–Da Fano cells allowing it to spread across the coverslip; then smoothly drag the mounted needle away from that side of the coverslip to the other end, creating a very thin gelatine smear with cells embedded in it. This can also be done with another coverslip, as in a blood smear. Set coverslips, as they are prepared, on an ice tray with moistened absorbent paper previously kept in a freezer. Immerse the prepared and hardened coverslips in a Columbia jar (Figure 2.14a) with ice-cold tap water.

- Transfer the coverslips to a Columbia jar with a 3% solution of cold silver nitrate solution ($AgNO_3$) for 5-10 minutes, and then set them side on a tray covered with cold water.
- Develop the coverslips exposing them to UV light or to a bright sunny window under chilled tap water for at least 20 minutes. Check impregnation; if is too faint leave it for another 10-20 minutes. Keep water cool at all times.
- Wash thoroughly in tap water and at least three times, then dehydrate in alcohol series (50-70-95-100-100% 5 min in each), clear in two changes of 10 minutes each of xylene and mount with Canada balsam.

This impregnation technique stains kinetosomes and argyrome (silverline system), a characteristic important in the identification of certain species. This technique keeps the shape of the cell, but does not stain cytoplasmic structures such as nuclei. It is a technique that does not allow a clear delimitation of kinetosomes (Figure 2.4.d, e) as the silver carbonate or protargol techniques do, so it is generally a complementary technique to them and always necessary if silverline systems have to be defined. Alternatively, Klein silver nitrate impregnation or "dry" silver nitrate can be used to study silverline systems, although that will not be discussed here (Klein 1926; Klein 1958; Foissner 1991).

2.4.2.3 Silver Proteinate/Protargol staining (Kirby 1945)

Silver proteinate or protargol, originally described by Kirby (1945) for an intestinal trichomonad and then generalized for other protozoa by Moskowitz (1950), is the silver staining methodology most widely used in the taxonomic study of ciliates. It is, however, a laborious protocol that requires some experience to be successful. Modifications to the original protocols have been published (see Foissner 1991 and literature cited therein), and the main variations include those to stain cells attached to a slide, or free cells, or even a quantitative modification that allows enumeration and identification at the same time (Montagnes and Lynn 1987). In our experience, the easiest and fastest method is the modification described by Wilbert (1975) which has been slightly modified by Song (personal communication and personal experience) to deal easily with large cells. This method is a free cell method and cells are processed manually with micropipettes in watch glasses or depression slides (Figure 2.3d, e):

- Fix cell in Bouin for 10 to 20 minutes maximum. The best method is to drop cells with a micropipette in a small volume of Bouin. Bouin is prepared with 15 parts of a saturated solution of picric acid, 5 parts of commercial formaldehyde (37%) and 1 part of glacial acetic acid (add just before use).

- Wash with distilled water, adding and removing water carefully in the same watch glass trying to keep cells concentrated in the centre. Repeat the process until the yellow colour of the Bouin fixative disappears completely.
- Concentrate cells in the centre of the watch glass with a very small quantity of water.
- Clear cells (until they become transparent; check under a low power microscope) with a ready made sodium hypochlorite solution (0.5% NaClO). This is a critical step; use 0.5% NaClO solution for large cells and for small cells dilute this hypochlorite solution one to four times with distilled water. Add small drops of this solution directly on top of the cells and leave 3-4 minutes or until cells are transparent.
- Wash several times with distilled water. This step is also very important since small traces of sodium hypochlorite solution (0.5% NaClO) will inhibit the next steps. Finally, leave cells in a small volume of water (aprox. 0.2 ml).
- Take a pinch of protargol (Roques or Merck) and sprinkle over the water volume. Seal the top of the watch glass with vaseline and place it on a hot plate at 60ºC for 40-50 minutes.
- Remove the protargol solution and develop cells directly dropping a 0.5% solution of hydroquinone (just before use, prepare 1% hydroquinone in a solution of 2.5% sodium sulfite and dilute as needed) and check staining under the microscope. Stop reaction with a 5% sodium thyosulphate solution ($Na_2S_2O_3$).
- Wash cells thoroughly with distilled water. Cells can be directly observed at this point and microscopic digital photograph documentation can be made. However, for permanent slides mix correctly stained cells with a very small drop of Mayer´s albumen (1:1 glycerol/ egg albumen and a small thymol crystal for preservation); eliminate excess albumen and dry on a hot plate.
- Dehydrate in alcohol series (50-70-90-95-98-100-100%) five minutes each, clear in two steps of xylene of 5 minutes each and mount with a synthetic mounting medium.

Other protargol methods start attaching fixed cells onto a slide, and then process the slide through the reagents in staining jars (Figure 2.14c). These methodologies are more difficult to carry out successfully although they produce better permanent records.

This methodology stains kinetosomes but no other infraciliary structures (such as kinetodesmal fibers or other microtubular structures). It may stain cilia, nuclei, contractile vacuole pores, and occasionally very faintly other cytoskeletic structures such as nassa, etc. (Figure 2.4f, g).

2.4.3 Flutax staining of ciliates

This methodology has recently been proposed (Arregui *et al.* 2002; Arregui *et al.* 2003) as a simple and easy alternative to silver staining methodologies to identify ciliates. The procedure successfully shows the main infraciliature patterns and location of structures important for the taxonomic identification of ciliates (Figures 2.22–2.24). Flutax is a fluorescent derivative of taxol (Paclitaxel) that binds specifically to microtubules, and since the main structures with taxonomic value in ciliates are microtubular, this methodology is very valuable due to its simplicity and fast results. It can also be used as a vital staining since it does not always require previous fixation or permeabilization, although a permeabilization process might provide better results in some ciliates such as hypotrichs, normally difficult to stain with silver impregnations due to their fragility (Arregui *et al.* 2003).

This protocol involves briefly the concentration of ciliates and incubation with 1μM Flutax solution for 10 minutes. Cells are then set into a drop of antifading agent (Citifluor Mounting Media; Microscopy Mart, USA) and the observation can be made immediately under a coverslip in an epifluorescence microscope with a standard fluorescein filter (Arregui *et al.* 2002). In those ciliates with a thick cortex, a pre-treatment of 30 seconds with Triton-X100 (0.1% in PHEM microtubule-stabilizing buffer) might be necessary to permeabilize cells, allowing an efficient penetration of Flutax in the cell to obtain better results (Arregui *et al.* 2002). For the more fragile hyprotrichs, previous permeabilization can also be made with 0.5% saponin in PHEM buffer for 10-15 seconds, but in this case a posterior fixation with paraformaldehyde (2% in PHEM buffer) is necessary before adding FLUTAX (Arregui *et al.* 2003).

2.4 THE INDICATOR VALUE OF CILIATES

Although studies of ciliates in WWTP have always been focussed on their use as indicators of efficiency in the purification process, effluent quality or other operational parameters, in fact few works have provided a straightforward tool –a biotic index – for plant operators to evaluate directly and easily the state of the process (Curds and Cockburn 1970b; Madoni 1994a). Among the works devoted to provide a simple methodology, the publication by Curds and Cockburn (1970b) was one of the first; it described a method to enable an evaluation of the average effluent quality within four BOD ranges (very high quality, high quality, moderate quality, low quality) based on knowledge about the protozoa species –mainly ciliates – present in the biological process. The authors findings were based on a relative assessment of protozoa abundance – as large, moderate or small numbers (i.e. no quantifications were done) – and appearance frequencies in different plants (% of species appearance). Punctuations were then given to each ciliate, meaning the degree of association to a specific category of effluent quality. Predictions with this methodology were tested to be correct in 83% of the cases studied.

Madoni (1994a) proposed a sludge biotic index (SBI) to evaluate activated sludge plant performance. To assess this index, protozoa were classified into functional groups, or protozoan key groups, and their changes were related to the environmental and operational conditions of the plant. This work was based on relative and mean abundances giving rise to four effluent quality classes. This methodology has proven to be very useful in estimating the biological quality of the sludge with a numerical value, and it is simple enough to allow plant operators accessible ciliate identification on a daily basis.

Other types of index, such as saprobic indicators, have also been used (Luna-Pabello *et al.* 1996). The main problem with this methodology is that it requires careful and accurate species identification to be able to provide a specific saprobic value, and this requires experienced knowledge. Another problem, identified by Curds and Cockburn (1970b), is that most species in wastewater treatment plants might be mesosaprobic, and in many cases eurisaprobic (Luna-Pabello *et al.* 1996), and therefore this classification is not very useful to assess biological wastewater treatments.

Other works have also provided valuable information about the association of ciliates to certain parameters important to evaluate the efficiency of the system (Esteban *et al.* 1990, 1991, 1992; Al-Shahwani and Horan 1991; Salvadó 1994; Salvadó *et al.* 1995; Martín-Cereceda *et al.* 2002; Lee *et al.* 2004). Some of these works have used multivariate statistical approaches after following different plants at different times of the year and intending to produce predictive equations that would easily allow an interpretation about the state of the processes. With this in mind, Al-Shahwani and Horan (1991) published a method to predict effluent quality and operating conditions in activated sludge plants through multiple regression analysis. Other works in activated sludge (such as Salvadó 1994) have proposed models based on mean cellular retention time (MCRT) ranges, and its relationship to microbial populations to allow monitoring of plant operation through a microscopic examination. Also in activated sludge, Salvadó *et al.* (1995) associated physical-chemical parameter groups to ranges of ciliate abundance, determining the optimal range of each species that could be used as indicator of effluent quality. Lee *et al.* (2004), working in an activated sludge pilot plant, also proposed ciliate indicators of different physical-chemical or operational plant parameters.

In rotating biological contactors, Kinner *et al.* (1988) related groups of ciliates to BOD_5 ranges in a similar way to what Curds and Cockburn (1970a) did in activated sludge. Martín-Cereceda *et al.* (2001a, b, 2002), again for rotating biological contactors, proposed a list of species related to an optimal efficiency in organic matter removal, and different ciliates were proposed as reliable indicators in these biological treatments.

This has been just a brief review of some of the possible approaches to assessing the indicator value of ciliates for plant performance. Readers are advised to refer to the original publications for a more detail account of the methodologies.

3

Systematic key to ciliate groups

3.1 KEY TO CILIATE GROUPS

About 160 ciliate species have been reported in wastewater treatment plants. Species identification is, however, complicated and some experience is needed due to their small size and the diversity of morphological characteristics. The following key is proposed in order to simplify the identification of species within a determined taxon; the latest taxonomic modifications are considered in this scheme (Puytorac *et al.* 1987; Lee *et al.* 2000; Cavalier-Smith 2002; Haussman *et al.* 2003; Adl *et al.* 2005).

In an attempt to provide a straightforward tool, this key is based mainly on characters easily recognizable by examination of living specimens with bright field or contrast microscopy (phase or interference). The existence and/or location of the cytostome are used as starting characteristics.

ISBN: 9781843391715. Published by IWA Publishing, London, UK.

Note: Specific taxonomic levels have been avoided. However, in some cases, a higher level of classification has been included in brackets to indicate those related taxonomic groups.

(1)	With cytostome	2
	Without cytostome	18
(2)	With oral cavity	3
	Without oral cavity (superficial cytostome)	15
(3)	With an adoral zone of membranelles (AZM)	4
	With few (3-4) membranelles/peniculi at the left side of the oral cavity	8
(4)	Homogeneous somatic ciliature (usually **swimming ciliates**)	5
	Somatic cilia grouped in specialized structures	6
(5)	**HETEROTRICHIDS**	
	Uniformly ciliated body	
(5.1)	Elongated flattened cells	***Spirostomum***
(5.2)	Ellipsoidal cells with prominent paroral membrane	***Blepharisma***
(5.3)	Trumpeted or ovoid cell with anterior peristome	***Stentor***
(6)	**OLIGOTRICHIDS (SPIROTRICHS)**	
	Body lacking somatic cilia or with few somatic cilia (**swimming ciliates**)	
(6.1)	Ovoid cells with groups of equatorial long cilia	***Halteria***
	Ventral cirri and dorso-ventrally flattened body (**crawling ciliates**)	7
(7)	**HYPOTRICHS (SPIROTRICHS)**	
	One, two or none marginal cirri	
(7.1)	Fronto-ventral, transverse, caudal and one or two marginal cirri, anterior AZM	***Euplotes***
(7.2)	Frontal and transverse cirri, posterior AZM	***Aspidisca***

	STICHOTRICHS (SPIROTRICHS)	
	Marginal rows of cirri	
(7.3)	Groups of fronto-ventral, transverse and long caudal cirri, rows of marginal cirri, well developed anterior AZM	***Stylonychia***
(7.4)	Groups of fronto-ventral, transverse and caudal cirri, rows of marginal cirri, anterior AZM	***Oxytricha***
(7.5)	Groups of frontal, caudal and transverse cirri, rows of ventral and marginal cirri, anterior short AZM	***Uroleptus***

(8)	Homogeneous somatic ciliature	9
	Absence of somatic ciliature	10

(9)	**TETRAHYMENIDS (OLIGOHYMENOPHORES)**	
	Oral ciliature with three membranelles and one paroral kinety **(swimming ciliates)**	
(9.1)	Pyriform cells with an anterior small oral cavity	***Tetrahymena***
(9.2)	Oval cells with an anterior torsion	***Dexiostoma/ Colpidium***
(9.3)	Ovoid cells with a large oral cavity	***Glaucoma***

	PENICULIDS (OLIGOHYMENOPHORES)	
	Oral ciliature with three peniculi and one paroral kinety, subpellicular trichocysts **(swimming ciliates)**	
(9.4)	Elongated or foot-shaped ciliates with a prominent equatorial oral groove region	***Paramecium***
(9.5)	Ovoid flattened ciliate with an anterior or middle oral cavity	***Frontonia***

	SCUTICOCILIATES (OLIGOHYMENOPHORES)	
	Oral ciliature with three membranelles and a prominent paroral kinety, long somatic cilia and one/several caudal cilia, small size **(swimming ciliates)**	
(9.6)	Ovoid cell with a truncated apical end, prominent paroral membrane	***Cyclidium***
(9.7)	Ovoid cell with a truncated apical end, small paroral membrane	***Uronema***
(9.8)	Flattened ovoid cell with a posterior oral cavity	***Cinetochilum***

(10)	**PERITRICHS (OLIGOHYMENOPHORES)**	
	Oral ciliature with peristomial haplokinety and polykinety and three infundibular peniculi	
-	Swimming ciliates	11
-	Sessile ciliates	12

(11)	Free-swimming vegetative cells	
(11.1)	Bell-shaped cells	***Astylozoon***
(11.2)	Barrel-shaped ciliates	***Opisthonecta***

(12)	Sessile vegetative cells with free-swimming larval stages	
-	Single cells	13
-	Colonial	14

(13)	With a contractile stalk (spasmoneme present)	***Vorticella***
-	Without a contractile stalk, lorica without valve	***Vaginicola***
-	Without a contractile stalk, lorica with valve	***Thuricola***

(14)	With contractile stalks (spasmoneme present)	
(14.1)	Single cells contractility (discontinuous spasmoneme)	***Carchesium***
(14.2)	Colonial contractility (continuous spasmonema)	***Zoothamnium***

-	Without contractile stalks (spasmoneme absent)	
(14.3)	With a large peristomial lip	***Epistylis***
(14.4)	Without peristomial lip, with opercule	***Opercularia***

(15)	Ovoid (**swimming ciliates**)	16
	Flattened body (**crawling ciliates)**	17

(16)	**PRORODONTIDS (PROSTOMATES)**	
	Apical cytostome with cytopharyngeal apparatus, caudal cilia	
(16.1)	Barrel-shaped ciliates with regularly disposed calcareous plates	***Coleps***

	HAPTORIDS (LITOSTOMATES)	
	Apical or subapical cytostome, oral toxicysts, regular somatic ciliature	
(16.3)	Oval cells with subapical cytostome	***Enchelys***
(16.4)	Spindle shaped cells with a long flexible and contractile neck	***Lacrymaria***
(16.5)	Elongated/ovoid cells with a broad anterior oblique (expansible) cytostome	***Spathidium***
(16.6)	Elongated cells with an anterior proboscis and cytostome placed at its basis	***Dileptus***

(17)	**PLEUROSTOMATIDS (LITOSTOMATES)**	
	Body laterally compressed with different ciliature at left and right sides, apical or subapical cytostome	
	(swimming or temporary crawling ciliates)	
(17.1)	Lanceolated cells, laterally compressed	***Litonotus***
(17.2)	Small elongated cells with an anterior torsion	***Acineria***

	PHYLOPHARINGIDS (PHYLOPHARYNGES)	
	Dorso-ventrally compressed body, cytostome with a nassa, ventral somatic ciliature, prominent preoral kinety	
(17.3)	Comma-shaped cytopharingeal basket, continuous preoral kinety	***Chilodonella***
(17.4)	Short cytopharingeal basket, fragmented preoral kinety	***Pseudochilodonopsis***
(17.5)	Large cells with a straight cytopharingeal basket, continuous preoral kinety	***Trithigmostoma***
(17.6)	Small cells with a posterior spine, small cytopharingeal basket	***Trochilia***

(18)	**SUCTORIDS (PHYLOPHARYNGES)**	
	Sessile predatory ciliates without somatic ciliature, visible tentacles with toxicysts, ciliated larval stages	
(18.1)	Tentacles randomly distributed on the cell	***Podophrya***
(18.2)	Two anterior tufts of tentacles	***Tokophrya***
(18.3)	Two anterior prominences with tufts of tentacles	***Acineta***
(18.4)	With stalked lorica, several fascicles of anterior tentacles	***Metacineta***
(18.5)	Without lorica, stalked with anterior, lateral and posterior fascicles of tentacles	***Multifasciculatum***

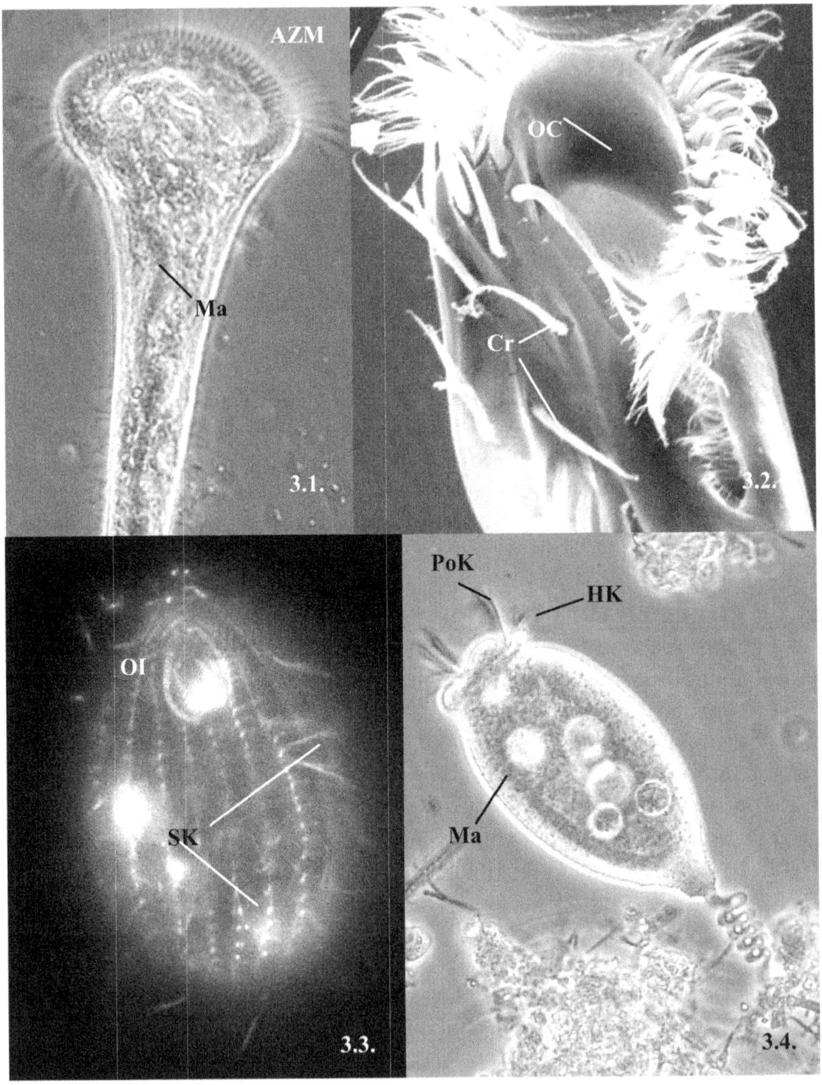

Figures 3.1–3.4 General aspects of the ciliate morphology. Figure 3.1 *Stentor roeseli*, phase contrast microscopy. Figure 3.2 *Euplotes* sp. SEM. Figure 3.3 *Tetrahymena thermophila*, Flutax stain. Figure 3.4 *Vorticella convallaria* complex, phase contrast microscopy. (AZM) adoral zone of membranelles; (Ma) macronucleus; (OC) oral cavity; (Cr) cirri; (OI) oral infraciliature; (SK) somatic kineties; (PoK) polikinety; (HK) haplokinety.

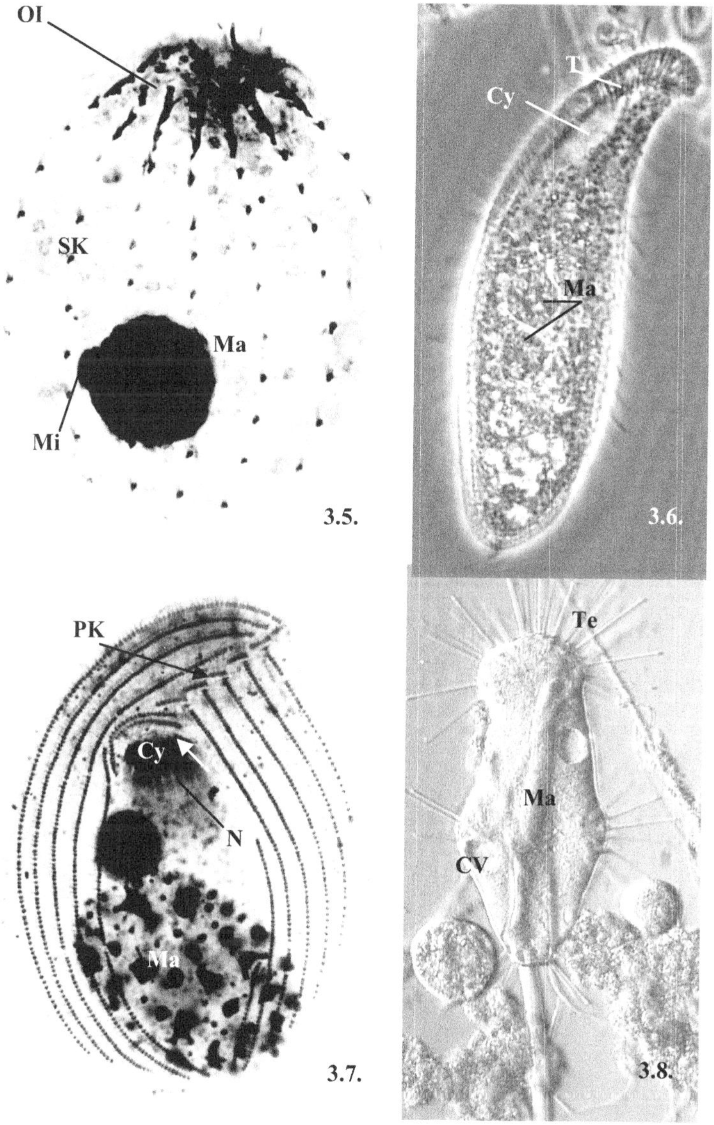

Figures 3.5–3.8 General aspects of the ciliate morphology. Figure 3.5 *Coleps hirtus*, silver carbonate impregnation (Fernández-Galiano 1976). Figure 3.6 *Litonotus lamella*, phase contrast microscopy. Figure 3.7 *Trithigmostoma cucullulus*, silver carbonate impregnation (Fernández-Galiano 1976). Figure 3.8 *Multifasciculatum elegans*, phase contrast microscopy. (Mi) micronucleus; (Ma) macronucleus; (T) toxicysts; (Cy) cytostome; (OI) oral infraciliature; (SK) somatic kineties; (PK) preoral kinety; (N) nassa; (CV) contractile vacuole; (Te) tentacles.

4

Guidelines for the identification of species in wastewater treatment plants

The main species of ciliate communities in wastewater biological processes are illustrated in this chapter, ordered by higher taxonomic rank/group. Species descriptions have been included using morphological characteristics easily discernible using optical microscopy (phase or interference contrast is recommended) and in some cases through silver impregnation staining. Details about the physical-chemical conditions associated with the appearance of certain species in WWTP are also included, and referred to the original records. The specialized literature must be referred to for more detailed information about specific identifications (Dragesco 1966, 1970; Bick 1972; Corliss 1979; Guhl 1979; Wu and Curds 1979; Curds and Wu 1983; Foissner 1984; Warren 1986; Augustin and Foissner 1992; Foissner et al. 1991, 1992, 1994, 1995; Fernández-Galiano et al. 1996; http://www.nies.go.jp/chiiki1/protoz/; http://starcentral.mbl.edu/microscope; http://protist.i.hosei.ac.jp/PDB/Images/Protista.html). Note: scale bars in all microphotographs represent 20 micrometres. Average size has been provided.

ISBN: 9781843391715. Published by IWA Publishing, London, UK.

4.1 HETEROTRICHS

These ciliates are generally represented in sewage treatment plants by few species with low abundances. These are free swimming species and occasionally sessile forms. The cell body is very contractile due to the presence of longitudinal myonemes between somatic kineties.

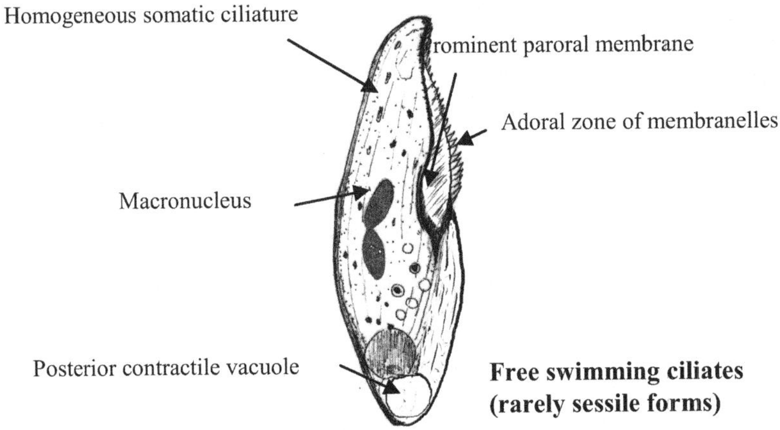

Figure 4.1. General characteristics of heterotrichs.

Cells elongated with a large terminal contractile vacuole: ***Spirostomum***

Species: ***Spirostomum teres*** Claparède and Lachmann, 1858

Size: 400 x 200 µm. Flattened species with an ovoid macronucleus; numerous conspicuous longitudinal somatic kineties spiralized when the body is contracted; the AZM does not reach the equatorial zone. Cells show a sliding movement with frequent cell torsions.

This species has a low tolerance to ammonia being reported in WWTP with a not very concentrated influent; it can tolerate low levels of oxygen and it is a good bioindicator to evaluate the toxicity of waters polluted by heavy metals, pesticides and phenols (Bick 1972; Madoni 1988, 2000).

The species *Spirostomum ambiguum*, rarely reported in biological reactors, is very similar to *S. teres* but it has a moniliform macronucleus and 1mm in length.

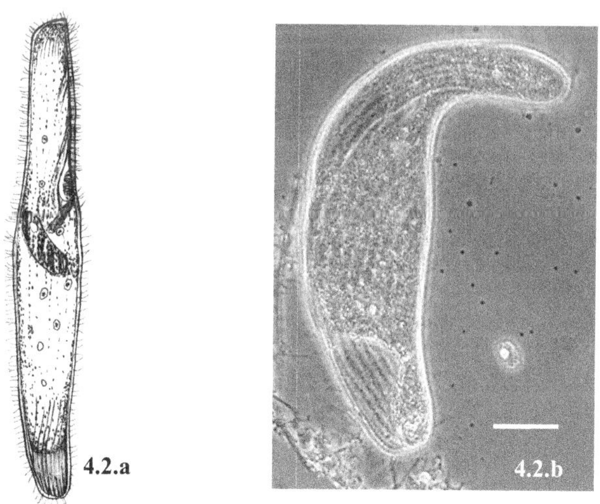

Figures 4.2.a and 4.2.b. ***Spirostomum teres*** Claparède and Lachmann, 1858.

Cells ellipsoidal or pyriform with tapered anterior end and rounded posterior end: *Blepharisma*

Species: ***Blepharisma undulans*** Stein, 1867

Size: 168 x 37 µm. Cells pink or red coloured with a large oral opening; oral ciliature with a prominent paroral membrane and AZM well developed exceeding the equatorial zone; cells present two macronuclear nodes and numerous longitudinal somatic kineties. Free swimming ciliates with a slow and continuous movement.

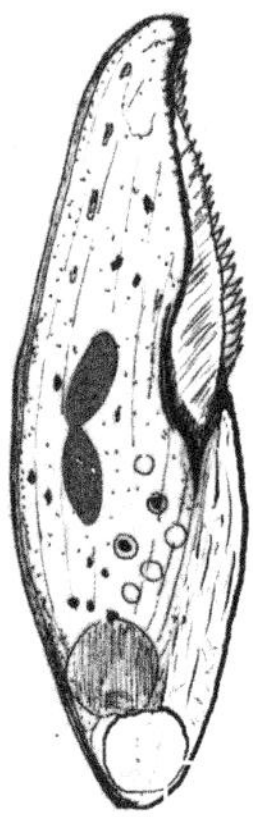

4.3.a

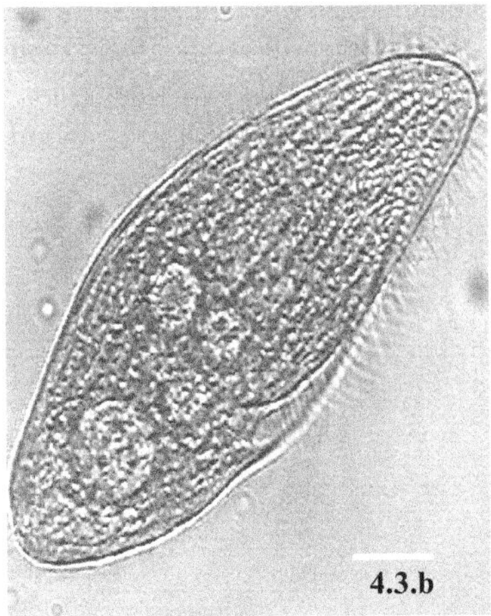

Figures 4.3.a and 4.3.b. ***Blepharisma undulans*** Stein, 1867.

Species: ***Blepharisma americanum*** Suzuki, 1954

Size: 179 x 95 µm. Cells pink coloured; oral ciliature with a prominent paroral membrane, the conspicuous AZM does not reach the equatorial zone; cells present several macronuclear nodes which can be used to distinguish this species from *B. undulans*.

Blepharisma sp. has been associated with nitrifying plants (Poole 1984).

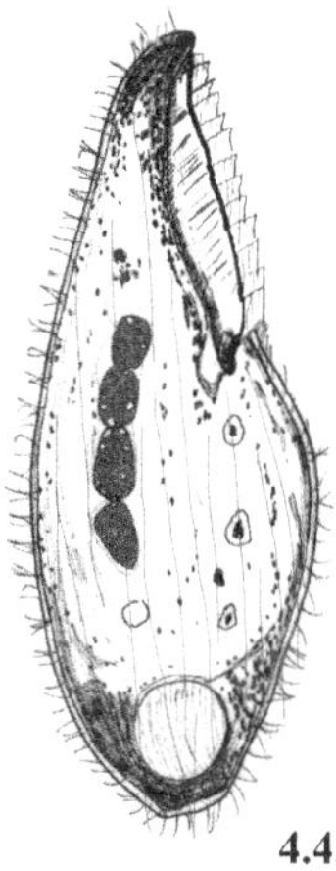

4.4.a

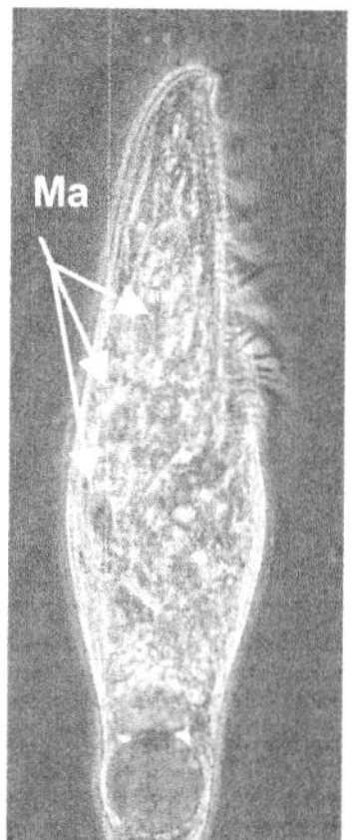

4.4.b

Figures 4.4.a and 4.4.b. ***Blepharisma americanum*** Suzuki, 1954.

Cells trumpet-shaped or ovoid, sessile or free-swimming. Apical peristome with a clear AZM. The body is contractile by the presence of myonemes between kineties: *Stentor*

Species: ***Stentor roeselii*** Ehrenberg, 1835

Size: 750 x 200 μm. Cells with elongated macronucleus and anterior contractile vacuole; often with a small lorica in the posterior pole. This species swims freely in the mixed liquor (ovoid cell shape) or sessile (trumpet-shaped cell).

Rarely reported in activated sludge, appears with low organic load and high sludge age (Madoni 1988; Isac *et al.* 2006).

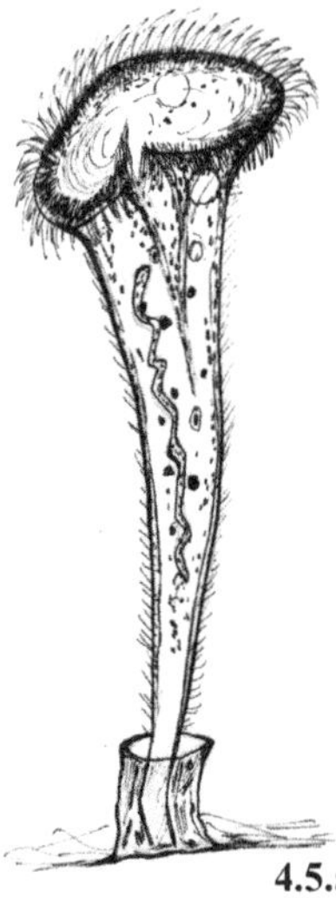

4.5.a

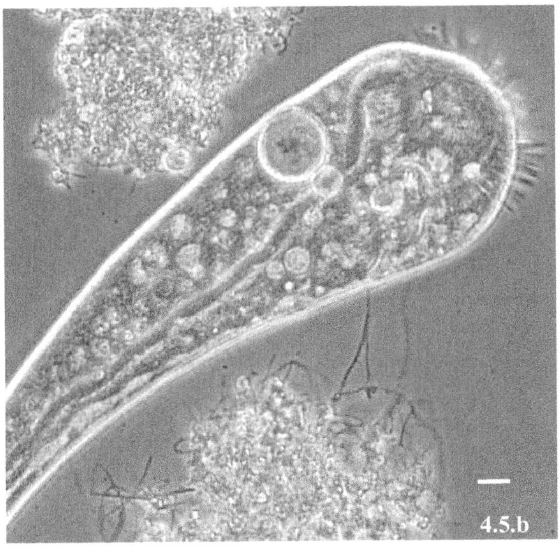

Figures 4.5.a and 4.5.b. ***Stentor roeselii*** Ehrenberg, 1835.

4.2 OLIGOTRICHIDS

This group is rarely reported in WWTP and just one species is generally found. In freshwater, these ciliates are free swimming with a characteristic "jumping" movement.

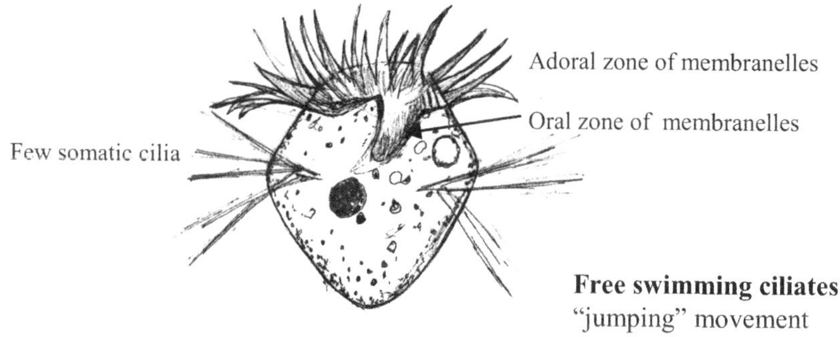

Figure 4.6. General characteristics of oligotrichids.

Cells with several equatorial groups of somatic cilia: ***Halteria***

Species: ***Halteria grandinella*** (Müller, 1773) Dujardin, 1841

Size: 30 µm in width. Body shape spherical or ellipsoidal with equatorial groups of long somatic cilia; oral ciliature with conspicuous membranelles in an anterior AZM. Swimming behaviour with a characteristic "jumping" alternated with slow rotatory movements.

H. grandinella is an excellent bioindicator to evaluate the toxicity of waters and wastewaters polluted by heavy metals due to the high sensitivity to cadmium and lead of this species. It prefers low ammonium levels (Bick 1972; Madoni and Romeo 2006).

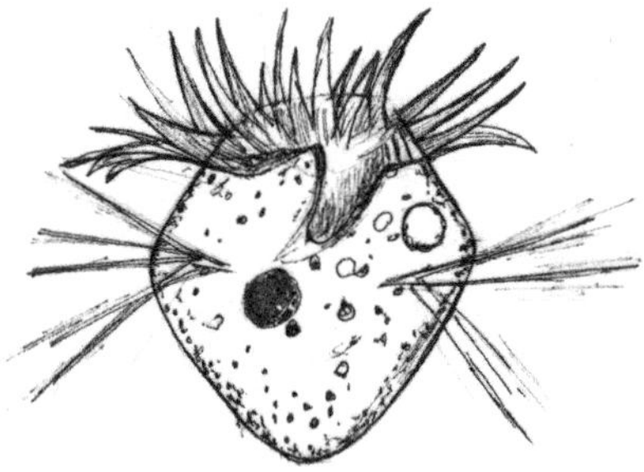

4.7.a

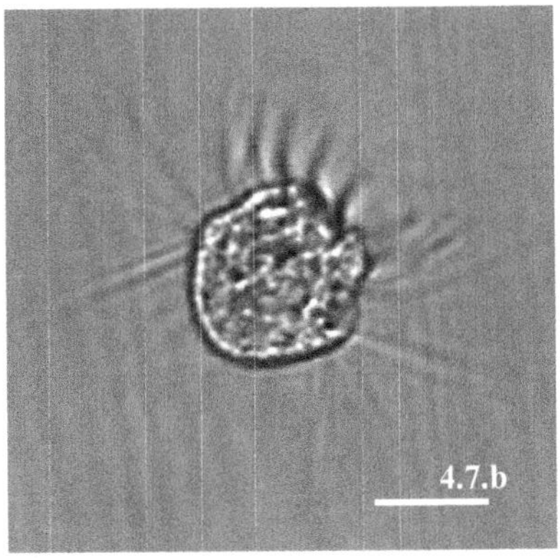

Figure 4.7.a and 4.7.b. ***Halteria grandinella*** (Müller, 1773) Dujardin, 1841.

4.3 HYPOTRICHS

This is one of the most representative groups of crawling ciliates. Several species usually present in high abundances in the biological treatment community.

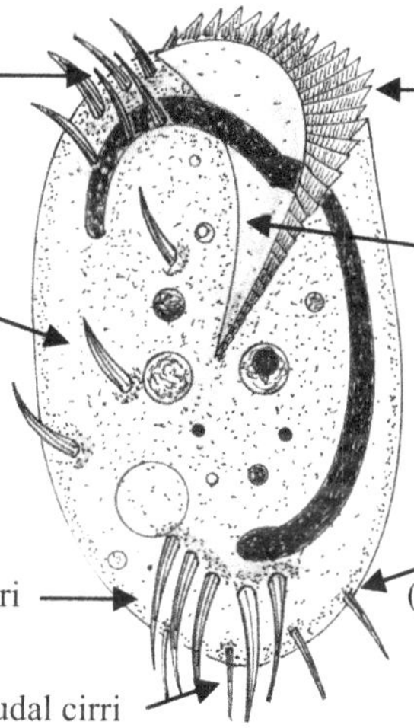

Figure 4.8. General characteristics of hypotrichs.

Ciliates with a well developed anterior oral opening and AZM. No rows of marginal cirri. Long fronto-ventral and transverse cirri. Short caudal and marginal cirri (0-2): *Euplotes*

Species: ***Euplotes aediculatus*** Pierson, 1943

Size: 135 x 85 μm. Flattened ciliate without patent dorsal ridges; well developed oral cavity with an inverted triangular overture; conspicuous transverse and caudal cirri; C-shaped longitudinal macronucleus. Ciliate algivorous, bacterivorous or predating on other protists. These ciliates crawl on the floc surface, moving its ventral cirri as independent units (when it is attached crawling on the floc the convex dorsal side can be observed) or swimming freely between the flocs.

E. aediculatus is an excellent bioindicator for evaluating the toxicity of waters and wastewaters polluted by nickel (Madoni and Romeo 2006).

E. patella presents similar morphological characteristics to those described for *E. aediculatus* but it has 5-7 easily distinguished dorsal ridges. This ciliate is associated with low organic loads (Madoni 1988).

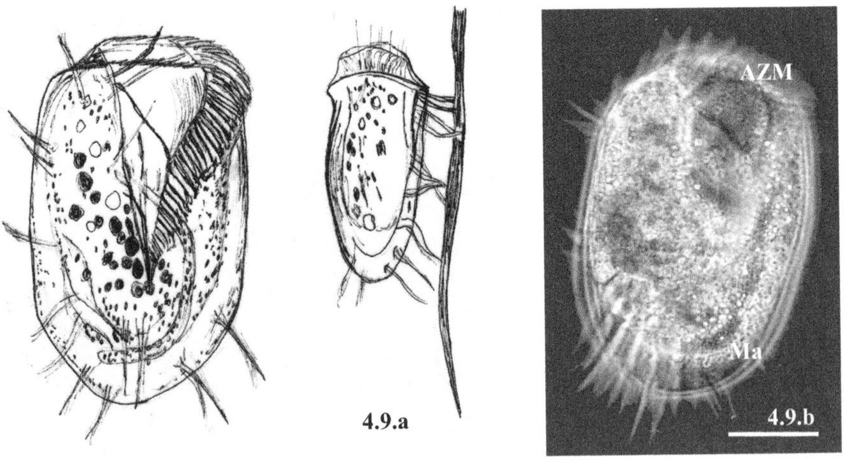

Figures 4.9.a and 4.9.b. ***Euplotes aediculatus*** Pierson, 1943.

Species: ***Euplotes affinis*** Dujardin, 1842

Size: 40-70 µm. Oval specimens with conspicuous dorsal ridges; longitudinal macronucleus C-shaped; well developed AZM. Small bacterivorous species of *Euplotes*.

This species is frequent in the colonization phase in activated sludge plants (Madoni 1988). Its occurrence in RBC systems is related to the adequate performance of the biological treatment in terms of organic matter removal (Martín-Cereceda *et al.* 2002).

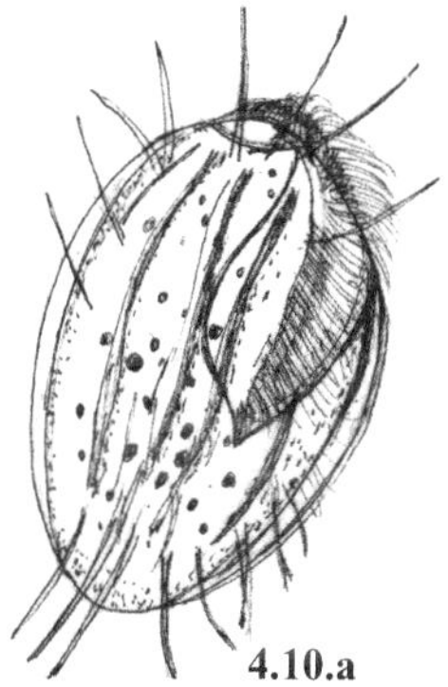

Figures 4.10.a and 4.10.b. ***Euplotes affinis*** Dujardin, 1842.

Species with a small AZM and long frontal and transverse cirri: ***Aspidisca***

Species: ***Aspidisca cicada*** *(A. costata)* (Müller, 1786) Claparède and Lachmann, 1858

Size: 32.5 x 30 μm. Rounded and flattened species with patent dorsal ridges; AZM posterior and C-shaped macronucleus. It is usually crawling on the floc surface on its ventral side.

This ciliate occurs commonly over a wide range of operating conditions and effluent quality (Poole 1984), although later on, *A. cicada* has been associated with stability conditions in activated sludge plants, frequently associated to *Vorticella convallaria* (Madoni 1988; Martín-Cereceda *et al.* 1996) and to old sludge age (Lee *et al.* 2004). This species also shows a remarkable tolerance to low concentrations of copper (Nicolau *et al.* 2001).

Ventral Side **Dorsal Side**

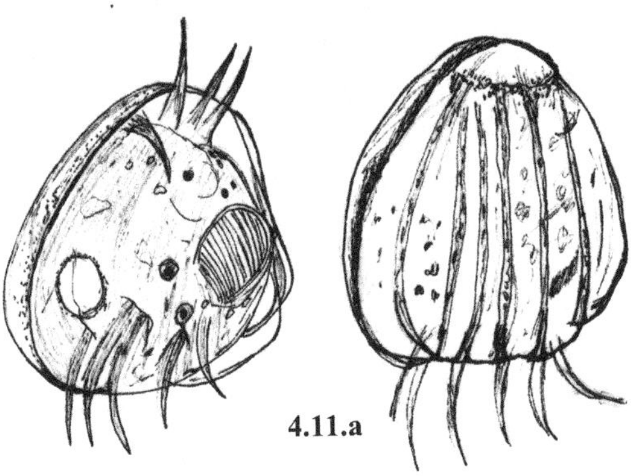

4.11.a

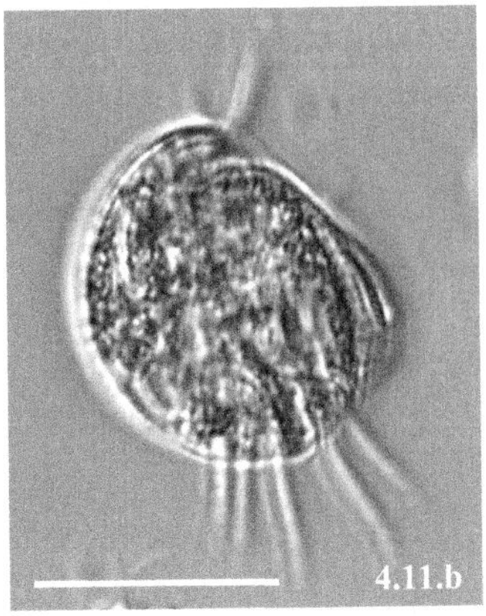

Figures 4.11.a and 4.11.b. ***Aspidisca cicada*** *(A. costata)* (Müller, 1786) Claparède and Lachmann, 1858.

Species: ***Aspidisca lynceus*** (Müller, 1773) Ehrenberg, 1830

Size: 42.5 x 37.5 µm. Species with a smooth dorsal side without ridges and a short distinct citoplasmic spine between transverse cirri. This bacterivorous ciliate moves over the flocs or biofilms.

This protozoa is rarely dominant in activated sludge plants. It is proposed to indicate sufficient aeration and in-depth treatment of effluents with good nitrification. Its presence appears to be incompatible with residual ammonia concentrations greater than 10-15 mg/l (Duchene 1991). Its occurrence in RBC systems is related to the adequate performance of the biological treatment in terms of organic matter removal (Martín-Cereceda *et al.* 2002).

4.12.a

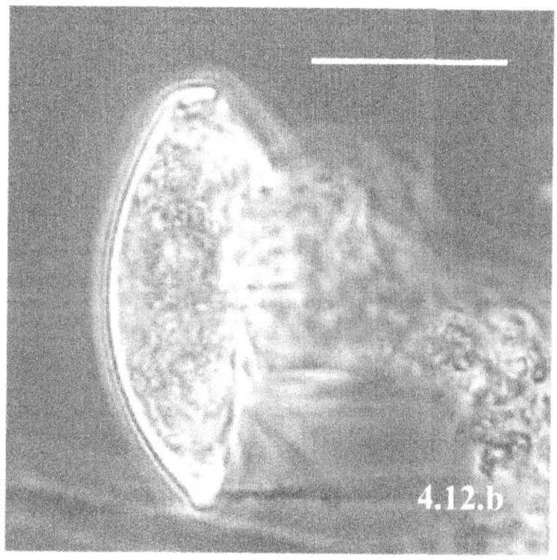

Figures 4.12.a and 4.12.b. ***Aspidisca lynceus*** (Müller, 1773) Ehrenberg, 1830.

Aspidisca turrita (*syn. Aspidisca lynceus*) (Ehrenberg, 1831) Claparède and Lachmann, 1858

Aspidisca with a prominent dorsal spine, differentiated when potential predators are present (Wicklow 1997).

4.4 STICHOTRICHS

Common species in activated sludge or biofilms WWTP, but with low abundances. Crawling ciliates associated to flocs or biofilms. Staining of the cells is recommended to determine cirral pattern of the species.

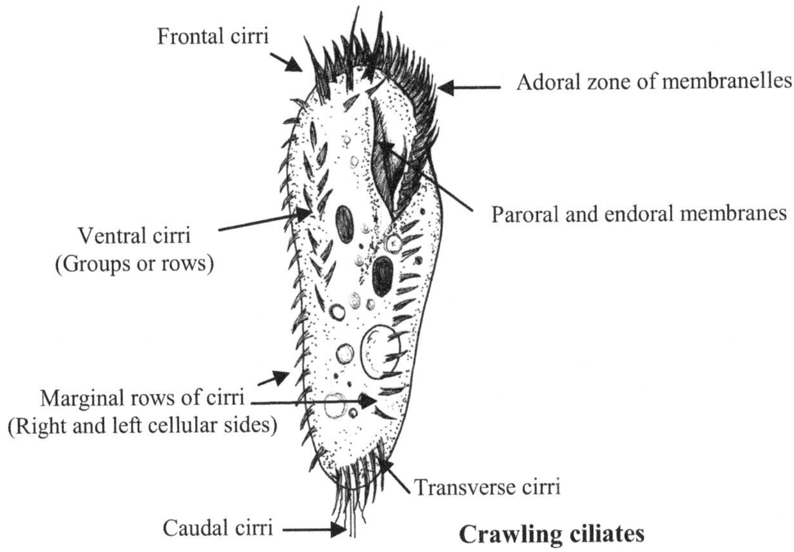

Figure 4.13. General characteristics of stichotrichs.

Ciliates with three long caudal cirri with a "trident" arrangement. Anterior AZM reaching the middle ventral side: ***Stylonychia***

Species: ***Stylonychia mytilus*** (Müller, 1773) Ehrenberg, 1830

Size: 300 x 70 µm. Large cells with anterior pole wider than the posterior one (triangular body shape); conspicuous AZM extending from the anterior part along the left side of the wide oral area; left and right rows of marginal cirri do not reach the posterior end of the cell. This ciliate can crawl on the surfaces or free swims in the mixed liquor (slow moving with frequent changes of direction). Predator on algae and other protists.

This is an uncommon species in sewage treatment plants such as activated sludge or trickling filters. However, *S. mytilus* has been described as more abundant in those plants treating dairy wastes compared to those plants treating municipal sewage (Berger 1999).

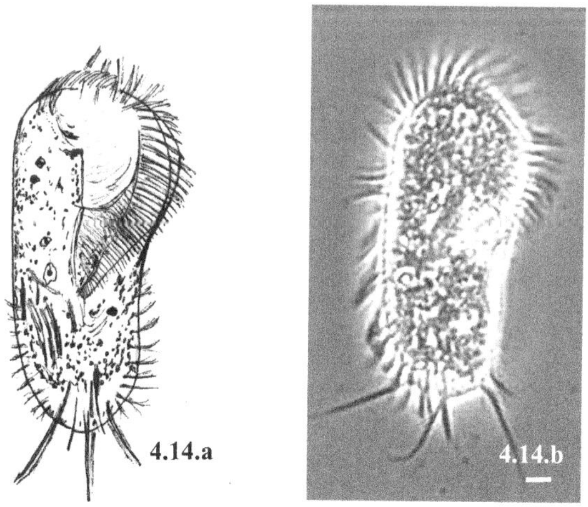

Figures 4.14.a and 4.14.b. ***Stylonychia mytilus*** (Müller, 1773) Ehrenberg, 1830.

S. putrina is a very similar species but smaller in size (135 µm) and with a narrower anterior pole.

Stichotrichs with three caudal cirri slightly longer than transverse ones. Anterior AZM reaching the anterior third part of the ventral side: *Oxytricha*

Species: ***Oxytricha fallax*** Stein, 1859

Size: 150 x 60 μm. Species ovoid and flexible with two oval macronuclei; left contractile vacuole near the posterior end of the AZM; left and right marginal rows of cirri reach the posterior end; caudal cilia shorter than those of the genus *Stylonychia*.

O. fallax has usually been recorded during maturation (second to fifth week) of the sludge (Berger 1999).

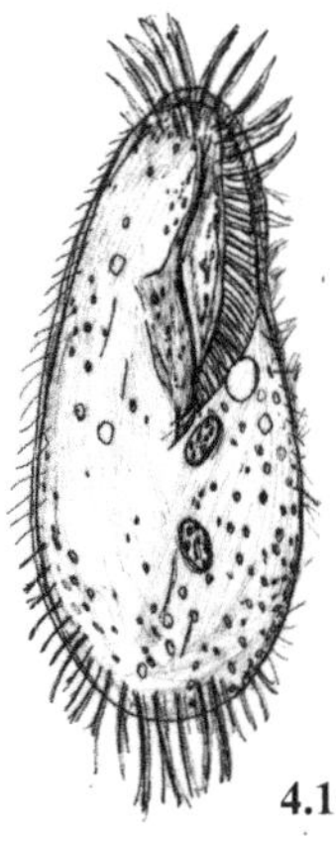

4.15.a

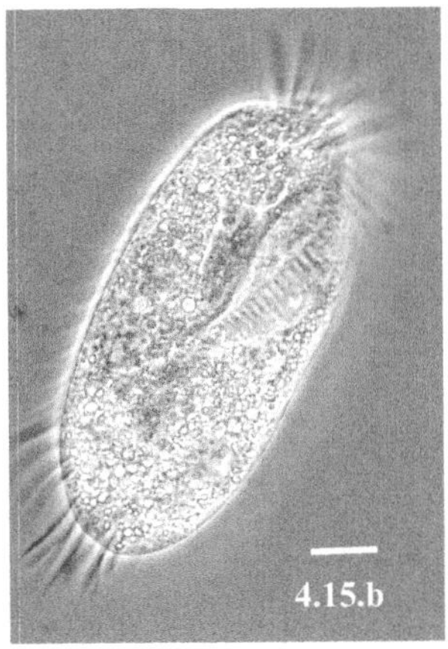

Figures 4.15.a and 4.15.b ***Oxytricha fallax*** Stein, 1859.

Species with a flexible body and a small AZM. Cirral pattern: rows of fronto-ventral and marginal cirri, J shaped transverse cirri: ***Uroleptus***

Species: ***Uroleptus limnetis*** Stokes, 1885

Size: 180 x 35 µm. Species with a tapered posterior end; three rows of fronto-ventral small cirri and three long caudal cirri; oral overture anterior and narrow. Crawling or free-swimming ciliates.

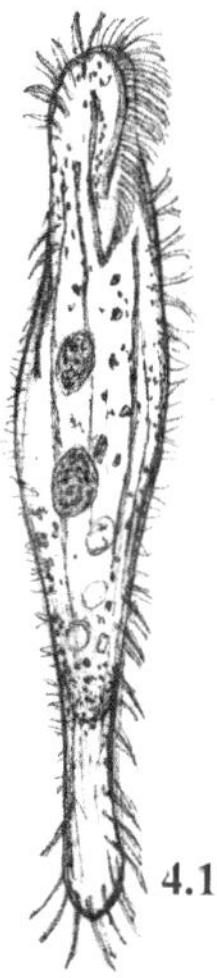
4.16.a

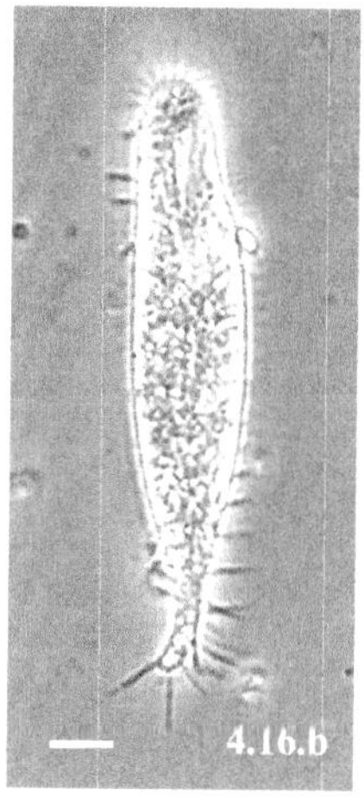

Figures 4.16.a and 4.16.b. ***Uroleptus limnetis*** Stokes, 1885.

4.5 TETRAHYMENIDS

These are free swimming ciliates. Live identification of some species in this group is difficult, so staining procedures are recommended.

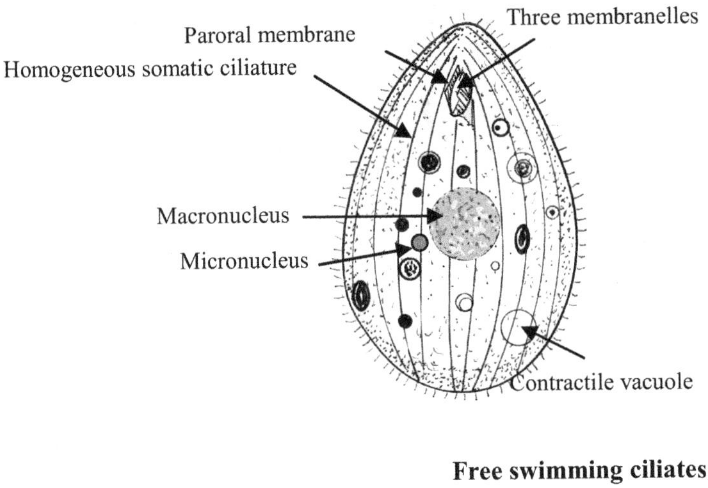

Figure 4.17. General characteristics of tetrahymenids.

Pyriform swimming cells: ***Tetrahymena***

Species: ***Tetrahymena pyriformis*** (Ehrenberg, 1830) Lwoff, 1947

Size: 65 x 55 µm. Pyriform cells with a small anterior oral overture; straight preoral suture resulting from the anterior convergence of the homogeneous somatic kineties; contractile vacuole in left posterior area; rounded macronucleus. Ciliate with a fast moving swimming behaviour.

This species may reach high abundances in biological reactors during colonization stages (developing activated sludge), and it is also related to low retention times (Madoni 1988). It was associated with high BOD_5 values within RBC systems wastewater treatment plants, generally in the first stages of the system (Luna-Pabello *et al.* 1996, Martín-Cereceda *et al.* 2002).

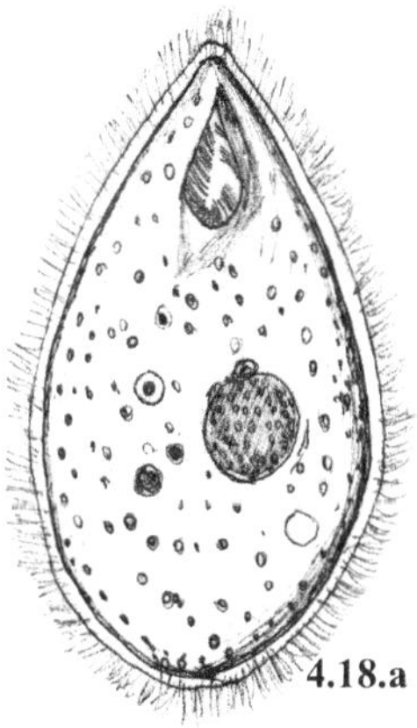

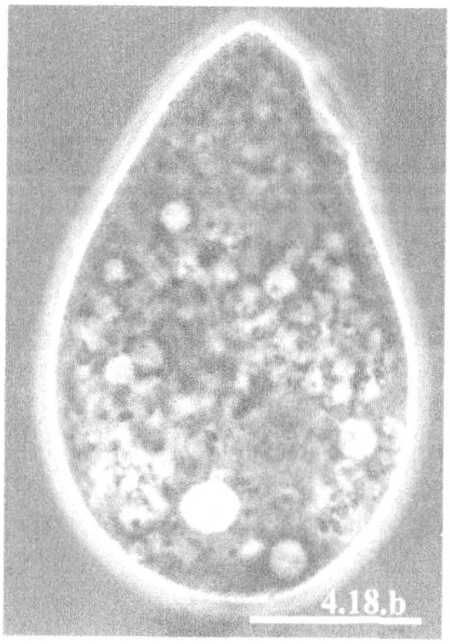

Figures 4.18.a and 4.18.b. ***Tetrahymena pyriformis*** (Ehrenberg, 1830) Lwoff, 1947.

Species: ***Tetrahymena thermophila*** Nanney and McCoy, 1976

Size: 60 x 40μm. Ovoid or pyriform species with numerous somatic kineties; anterior oral cavity; spherical macronucleus; posterior contractile vacuole.

The abundance of *T. thermophila* goes up when dissolved solids and BOD increase in biological reactors and with low water temperatures (Esteban *et al.* 1991, 1992).

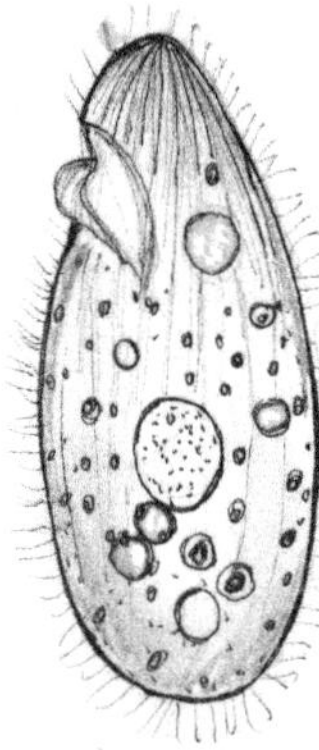
4.19.a

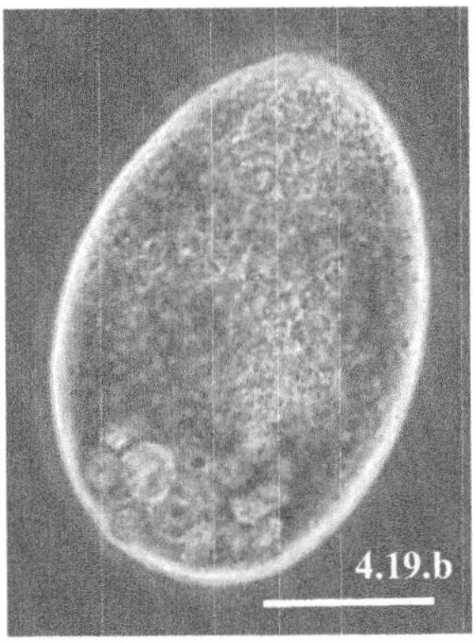

Figures 4.19.a and 4.19.b. ***Tetrahymena thermophila*** Nanney and McCoy, 1976.

Ovoid cells with a torsion at the anterior body end: *Dexiostoma/ Colpidium*

Species: ***Dexiostoma campyla*** (syn. *Colpidium campylum*) (Stokes, 1886) Jankowski, 1967

Size: 85 x 60 μm. Bean-shaped ciliate with a patent oral cavity just below the anterior torsion clearly observed in a lateral view; left posterior contractile vacuole; ovoid macronucleus. Fast movement in the mixed liquor.

Characteristic of activated sludge plants at starting phase (Madoni 1991). This ciliate resists the toxicity of the influent better than others (Madoni 1994a; Martín-Cereceda *et al.* 1996). Mostly encountered in the first stages within RBC system wastewater treatment plants where this species was associated with high BOD_5 values (Kinner *et al.* 1988; Luna-Pabello *et al.* 1990, 1996; Berri and Casaschi 1991; Curds 1993; Martín-Cereceda *et al.* 2002). *D. campyla* has also been described in wetlands treating primary effluent only in the inlet zone where higher amounts of organic matter are found (Puigagut *et al.* 2007).

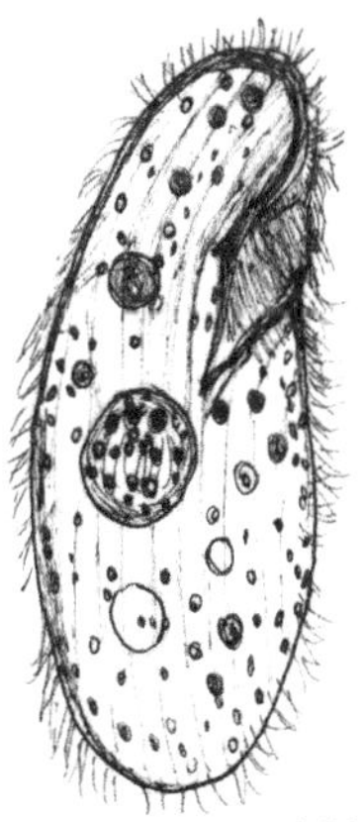
4.20.a

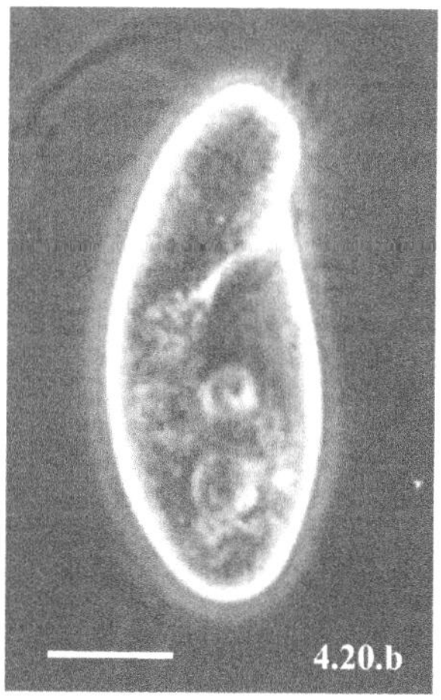

Figures 4.20.a and 4.20.b. ***Dexiostoma campyla*** (syn. *Colpidium campylum*) (Stokes, 1886) Jankowski, 1967.

Colpidium colpoda (Ehrenberg, 1831) Stein, 1869

Size: 130 x 120 μm. Species larger than *D. campyla*, but with similar morphological features; cells with a dense homogeneous somatic ciliature; oral cavity placed at the basis of the anterior body torsion (this one is less obvious than in *D. campyla*); left posterior contractile vacuole. These are free-swimming ciliates in the mixed liquor.

This species has also been encountered in the presence of high levels of organic matter (Cardinaletti and Zitelli 1991).

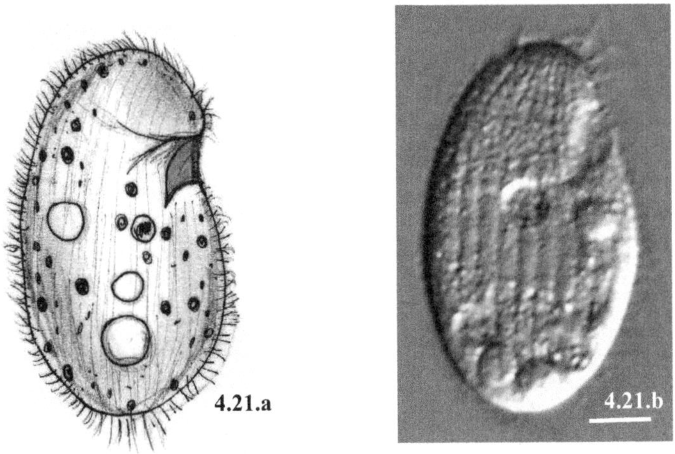

Figures 4.21.a and 4.21.b. ***Colpidium colpoda*** (Ehrenberg, 1831) Stein, 1869.

Large oral opening with patent oral ciliature: ***Glaucoma***

Species: ***Glaucoma scintillans*** Ehrenberg, 1830

Size: 60 x 55 μm. Ovoid or spherical and slightly flattened cells; oral cavity with prominent membranelles (left side) and paroral membrane (right side); one posterior contractile vacuole; oblique preoral suture. It can be observed free-swimming with a fast and straight movement.

This species, described in RBC WWTP, is very sensitive to small changes in inorganic physical-chemical parameters; it has been mostly encountered in the first stages of the system (Martín-Cereceda *et al.* 2002). It is rare in activated sludge plants. It seems to be highly tolerant to low oxygen concentration (Madoni 1988).

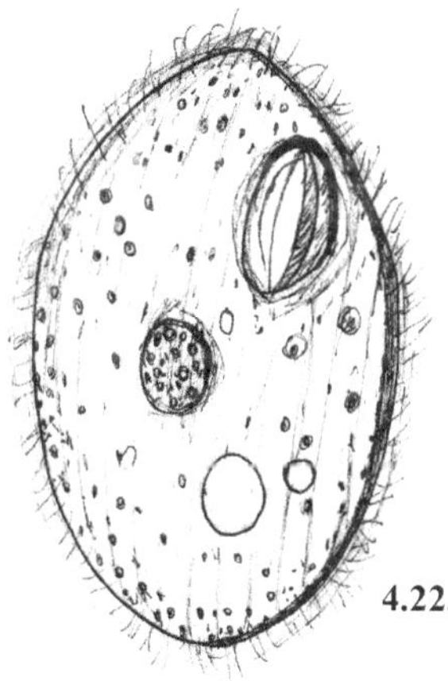

4.22.a

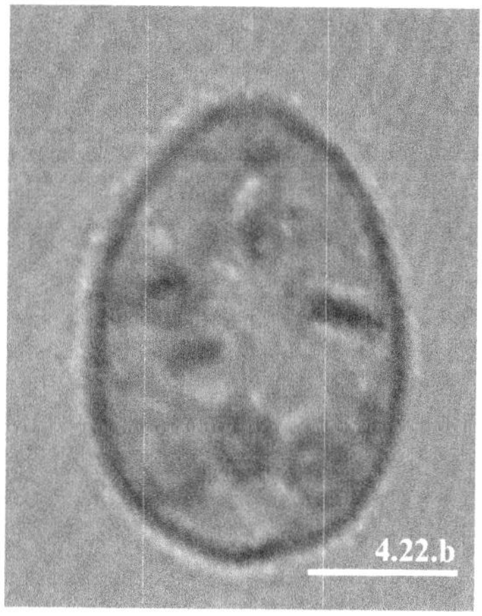

Figures 4.22.a and 4.22.b. ***Glaucoma scintillans*** Ehrenberg, 1830.

4.6 PENICULIDS

Fast swimming cells with low abundances in the WWTP, except in the first phases of colonization of the activated sludge plants (Madoni 1988, 1991).

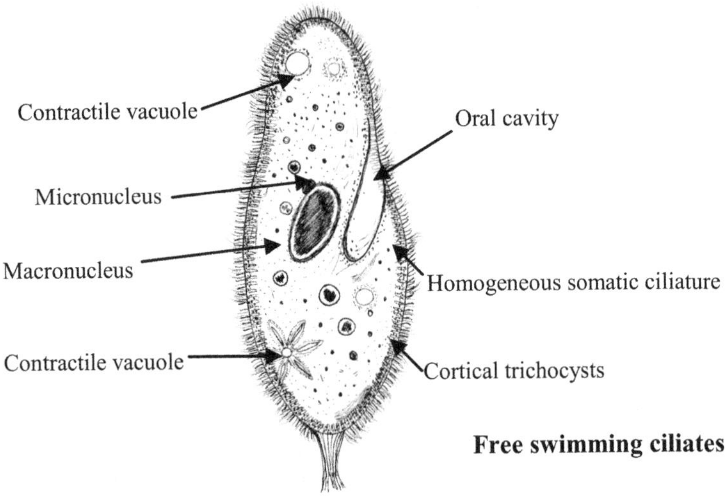

Figure 4.23. General characteristics of peniculids.

Oval or elongated foot-shaped cells with an equatorial torsion where the oral cavity is clearly distinguished. Numerous somatic kineties with a cluster of long caudal cilia. Subpellicular striation corresponding to the cortical location of trichocysts: *Paramecium*

Species: ***Paramecium aurelia*** Müller, 1773

Size: 180 x 100 μm. Rounded anterior and posterior ends; strongly patent oral torsion; two micronuclei and one macronucleus; two contractile vacuoles (anterior and posterior respectively); long caudal tuft of cilia. Fast moving cells.

Small paramecia have been encountered in high F/M values and sufficient oxygenation conditions (Duchene 1991).

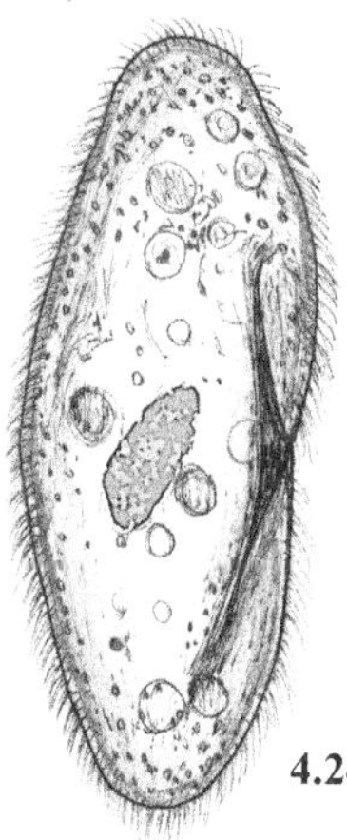

4.24.a

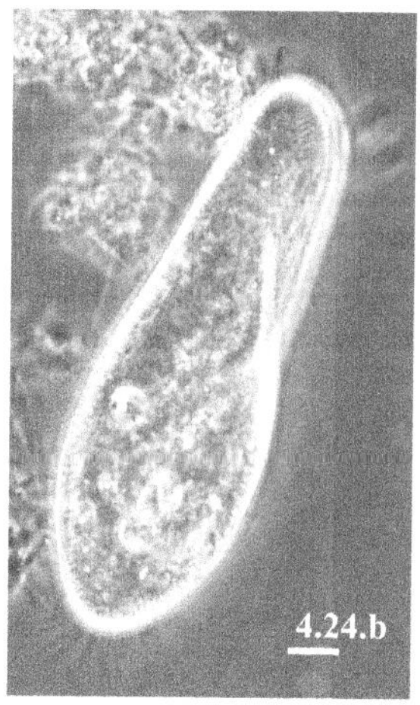

Figures 4.24.a and 4.24.b. ***Paramecium aurelia*** Müller, 1773.

Species: ***Paramecium bursaria*** (Ehrenberg, 1831) Focke, 1836

Size: 150 x 90 μm. Ovoid, slightly flattened cells with numerous endosymbiotic zoochlorellas; numerous somatic kineties around the cell and caudal cilia not very long; large oral overture slightly displaced from the middle to the anterior part of the body. Swimming freely in the mixed liquor of biological reactors.

This species tolerates a high concentration of ammonium and low levels of oxygen (Bick 1972).

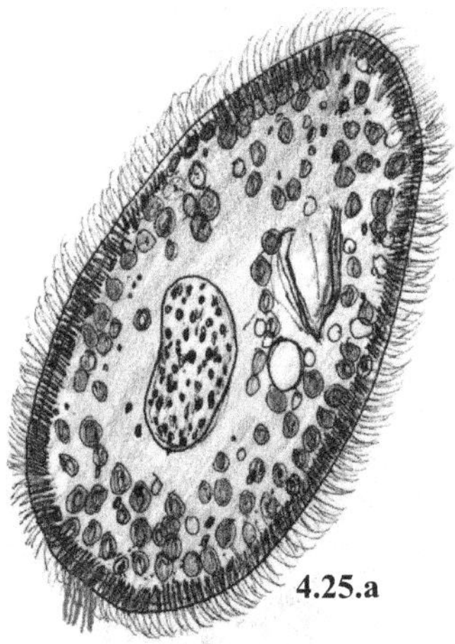

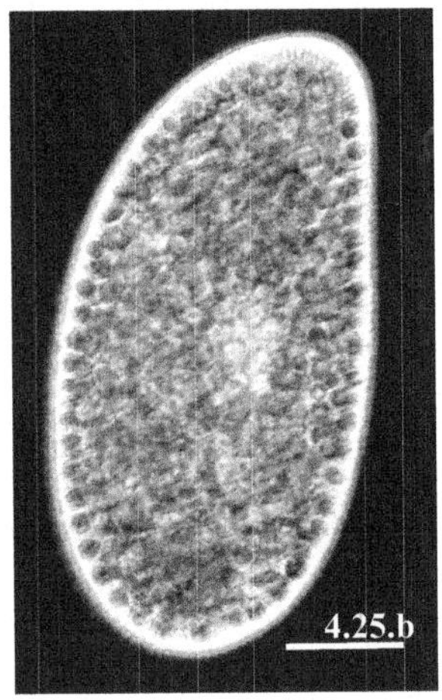

Figures 4.25.a and 4.25.b. ***Paramecium bursaria*** (Ehrenberg, 1831) Focke, 1836.

Species: ***Paramecium caudatum*** Ehrenberg, 1833

Size: 300 x 180 μm. Large cigar-like ciliate with an acute posterior end with a tuft of caudal cilia; one micronucleus and one macronucleus; long oral groove; two contractile vacuoles at both extremes of the cell. Fast movement ciliate.

Species detected rarely in activated sludge (Madoni 1988). Large paramecia disclose abundant oxygenation and good depuration treatment (Duchene 1991).

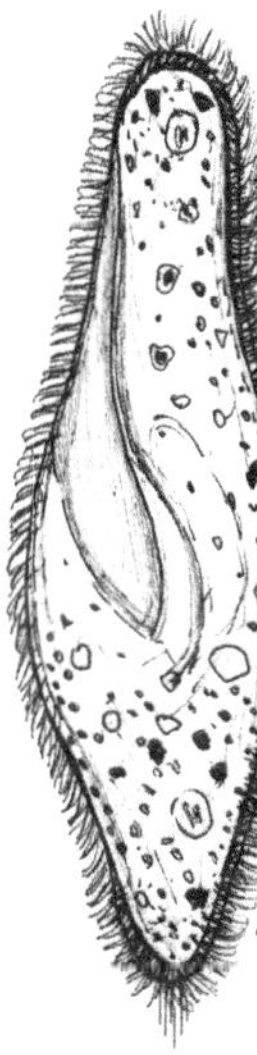

4.26.a

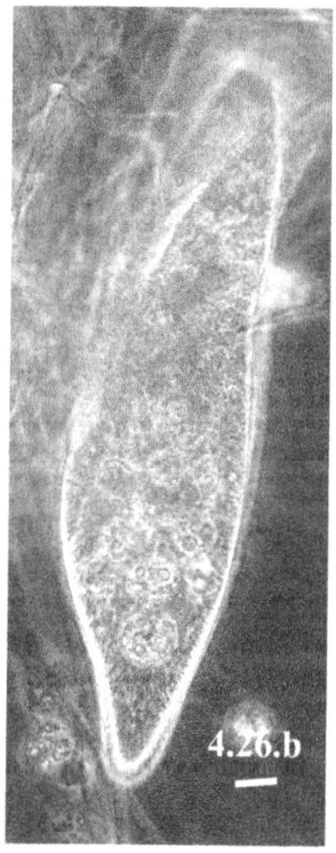

Figures 4.26.a and 4.26.b. ***Paramecium caudatum*** Ehrenberg, 1833.

Oval flattened cells with a conspicuous contractile vacuole. Anterior oral cavity and dense somatic ciliature: ***Frontonia***

Species: ***Frontonia leucas*** (Ehrenberg, 1833) Ehrenberg, 1838

Size: 150 x 300 μm. Large dorso-ventrally flattened ciliate; uniform somatic ciliature; one contractile vacuole with long patent radial canals; ovoid macronucleus. This species feeds on algae and testate amoebae. Slow straight movements.

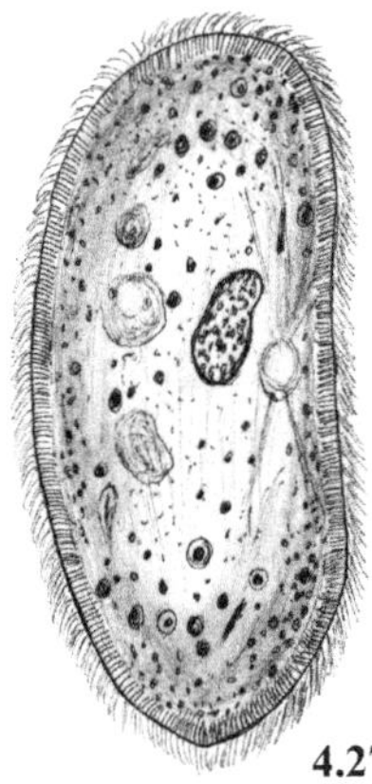

4.27.a

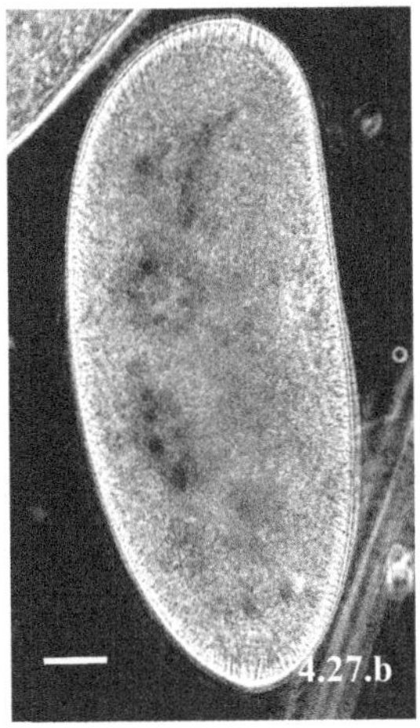

Figures 4.27.a and 4.27.b. ***Frontonia leucas*** (Ehrenberg, 1833) Ehrenberg, 1838.

4.7 SCUTICOCILIATES

Small cells with long somatic cilia, one or more caudal cilia are present. The paroral membrane is frequently patent at the right side of the oral cavity. Their movements are characteristic quick, jumpy with sudden stops. Ciliates very common in WWTP.

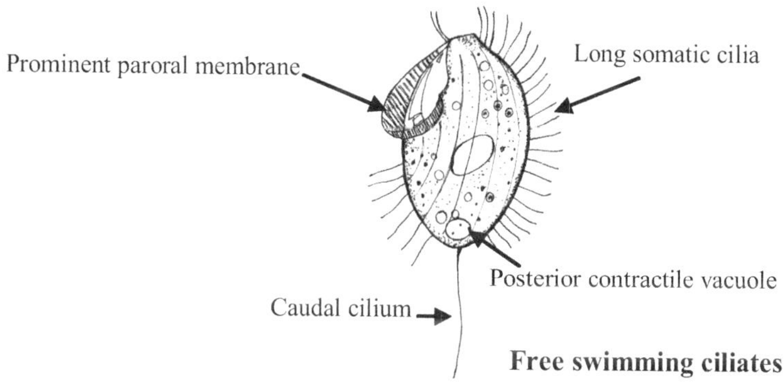

Figure 4.28. General characteristics of scuticociliates.

Cells with a large oral overture that reaches the half–body length. Truncate cilia-free anterior cap. The long paroral membrane is clearly visible when the ciliate stops to feed: ***Cyclidium***

Species: ***Cyclidium glaucoma*** Müller, 1773

Size: 20 x 9.5 µm. Oval flattened ciliate with one caudal cilium; large oral area with a prominent paroral membrane; posterior contractile vacuole. Fast moving alternates with sudden stops (when the cell stops the ciliature can be easily observed).

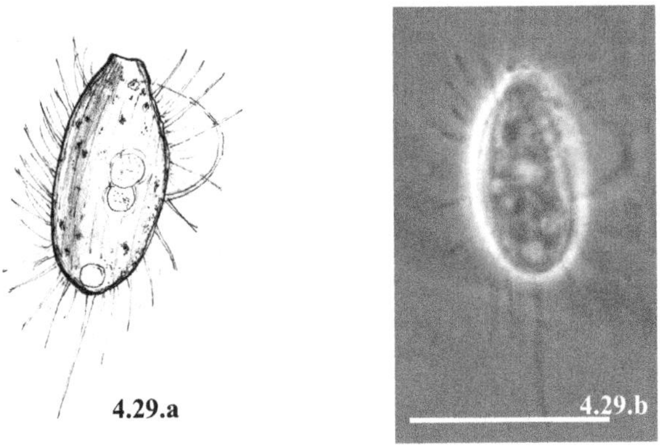

Figures 4.29.a and 4.29.b. ***Cyclidium glaucoma*** Müller, 1773.

Oval ciliates with a non ciliated anterior pole. Oral overture does not reach the half-body length. Paroral membrane less patent than in *Cyclidium*. One caudal cilium: ***Uronema***

Species: ***Uronema nigricans*** Müller, 1786

Size: 20 x 12 µm. Long somatic cilia, with a prominent caudal one; truncated anterior end; one posterior contractile vacuole. It moves with sudden movements, continuously changing its position.

This ciliate indicates low effluent quality (Salvadó *et al.* 1995).

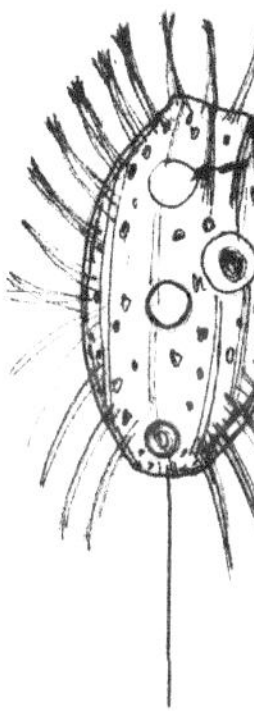

4.30.a

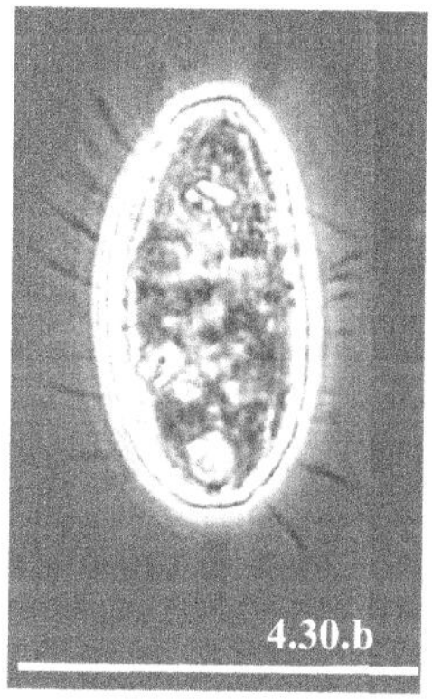

Figures 4.30.a and 4.30.b. ***Uronema nigricans*** Müller, 1786.

Dorso-ventrally flattened rounded ciliate. Posterior oral overture with a short AZM. Several caudal cilia. Subterminal contractile vacuole opposed to oral area: ***Cinetochilum***

Species: ***Cinetochilum margaritaceum*** (Ehrenberg, 1831) Perty, 1849

Size: 28 x 23.5 μm. Laterally flattened cells with patent cortical bending ridges in which somatic kineties are placed; tuft of long caudal cilia. It moves freely turning around.

Species observed in RBC systems; proposed as bioindicator of low efficiency of the RBC performance (Kinner *et al.* 1988; Curds 1993). It is a common species in activated sludge with good depuration efficiency, its tolerance to ammonium is low (Madoni 1988).

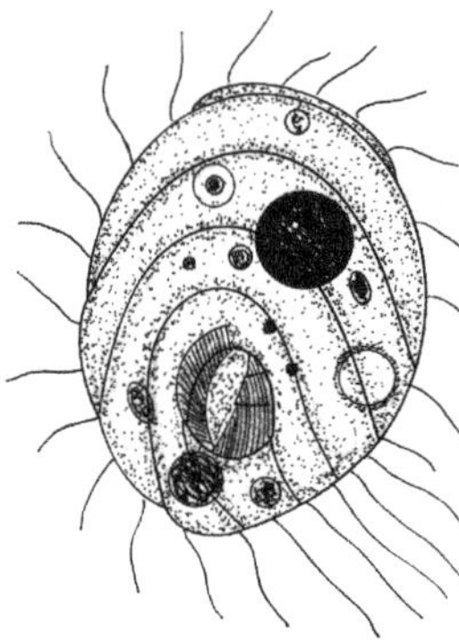
4.31.a

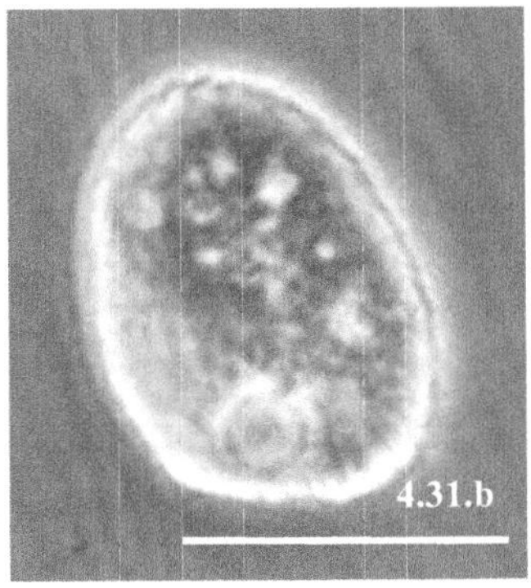

Figures 4.31.a and 4.31.b. ***Cinetochilum margaritaceum*** (Ehrenberg, 1831) Perty, 1849.

Ovoid dark ciliate with a long caudal cilium. Anterior truncate apex without cilia. Anterior oral overture with a slightly prominent paroral kinety: *Dexiotricha*

Species: ***Dexiotricha granulosa*** Kent, 1881

Ovoid, dark species with a great number of cytoplasmic granules; oral ciliature can be observed in a lateral view; somatic cilia are not as long as in other scuticociliates; one caudal cilium; posterior contractile vacuole. Ciliate with a homogeneous swimming movement.

Due to its size and morphology it can be easily confused with *Dexiostoma campyla*.

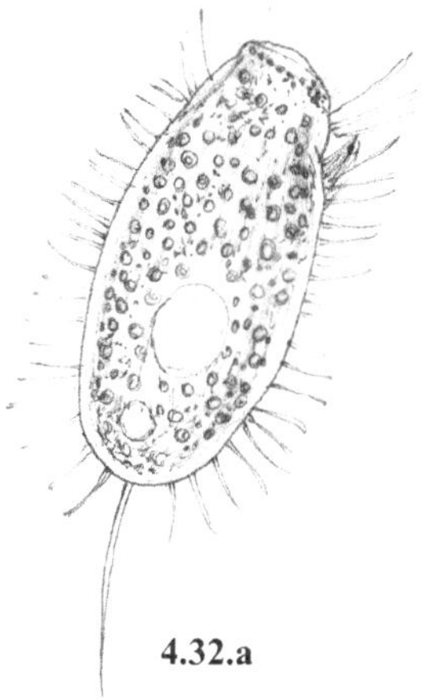

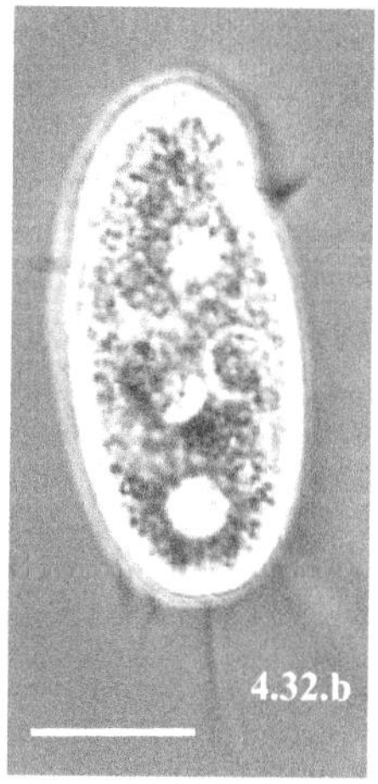

Figures 4.32.a and 4.32.b. ***Dexiotricha granulosa*** Kent, 1881.

4.8 PERITRICHS

The most characteristic group of ciliates associated to floc and biofilm in WWTP. The majority of the species are sessile although, under certain conditions, free swimming larval stages can be observed. Stalked specimens can be contractile (with spasmoneme) or not (without spasmoneme). Active bacterivorous. Species abundance is related to good activated sludge management (Martín-Cereceda *et al.* 1996).

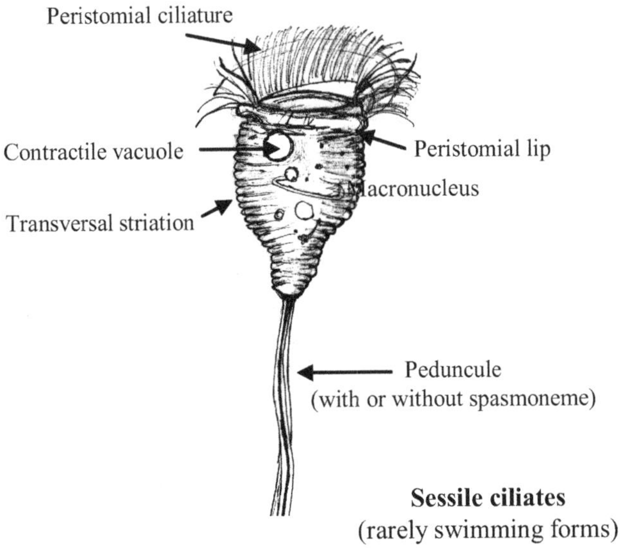

Figure 4.33. General characteristics of peritrichs.

Free-swimming ciliate. Somatic infraciliature is restricted to the aboral tuft of cilia: ***Astylozoon***

Species: ***Astylozoon fallax*** Engelmann, 1862

Size: 40 x 100 µm in length. Bell-shaped elongated cells with an easily observed transversal striation; anterior contractile vacuole; kidney-like macronucleus; acute posterior end with a tuft of cilia. Fast twister movement.

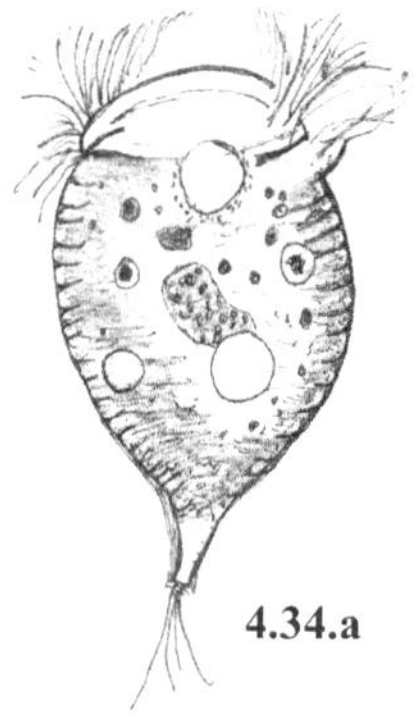

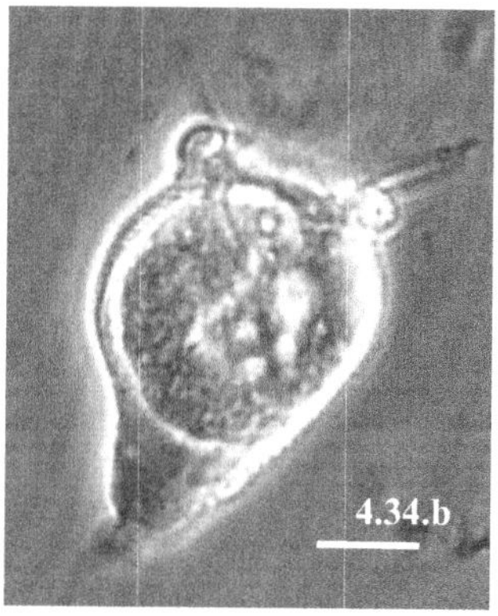

Figures 4.34.a and 4.34.b. ***Astylozoon fallax*** Engelmann, 1862.

Free-swimming cells without sessile stages. Barrel shape with a band of aboral ciliature clearly distinguished: *Opisthonecta*

Species: ***Opisthonecta henneguyi*** Fauré-Frèmiet, 1906

Size: 125 x 100 µm. Sac-like cells with a high number of transverse lines; single contractile vacuole; C-shaped transversal macronucleus; aboral band of somatic cilia. Free swimming ciliates, move with a rotary motion.

4.35.a

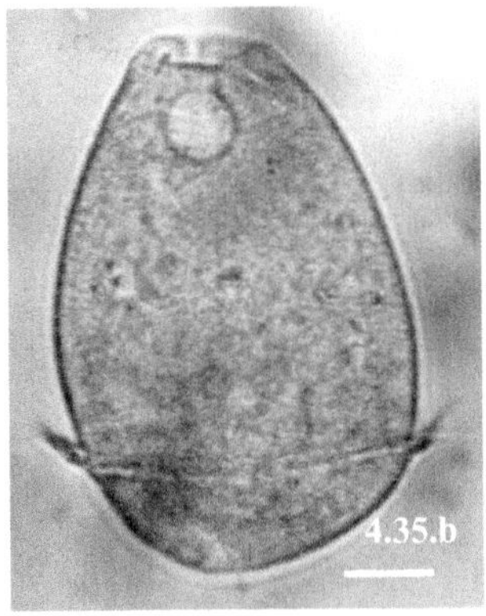

Figures 4.35.a and 4.35.b. ***Opisthonecta henneguyi*** Fauré-Frèmiet, 1906.

Sessile bell-like ciliates. Single zooids with a contractile peduncule (with internal spasmoneme): ***Vorticella***

Due to species identification difficulties, Foissner (1992) proposed the use of "complexes", the most significant in WWTP being the *Vorticella aquadulcis, Vorticella infusionum, Vorticella microstoma*, and *Vorticella convallaria* complexes.

Vorticella aquadulcis complex

Size: 35 x 20 µm. Zooid with less than 30 transverse striations (silver lines); macronucleus C-shaped and transversally placed; long stalk with sinusoid spasmoneme; one anterior contractile vacuole near the infundibulum. Cells associated to flocs or biofilms.

This complex includes the species *V. striata*, which is an indicator of poor quality effluent (Martín-Cereceda *et al.* 1996), although Madoni (1988) related the species with good oxygen conditions.

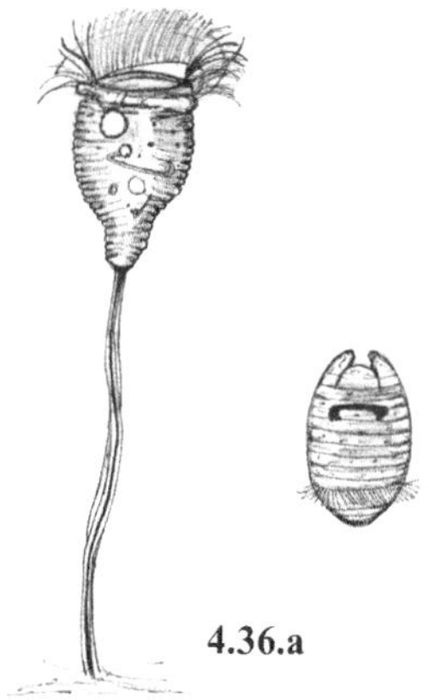

4.36.a

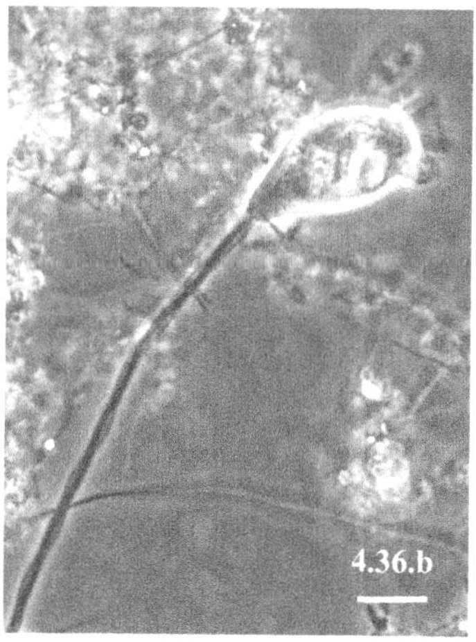

Figures 4.36.a and 4.36.b. ***Vorticella aquadulcis* complex**.

Vorticella infusionum complex

Size: 52 x 35 µm. Zooid with 30-45 transverse striations (silver lines); macronucleus C-shaped transversally placed; contractile vacuole anteriorly placed over the macronuleus; patent peristomial lip and peristome slightly narrower than the cellular width. Ciliates attached to biofilms or flocs.

High abundances occur in plants with elevated organic contents and low oxygen concentrations (Isac *et al.* 2006).

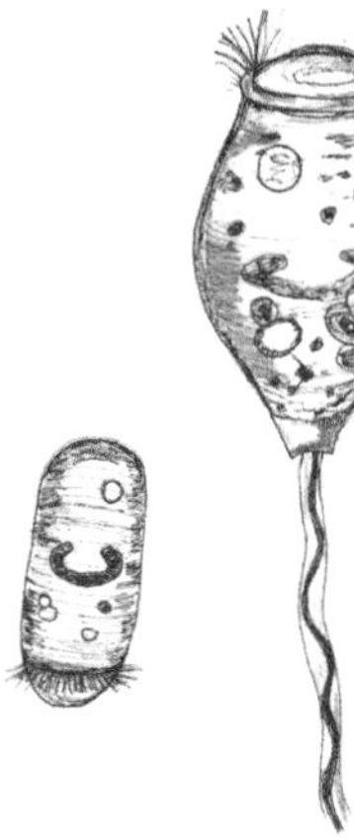

4.37.a

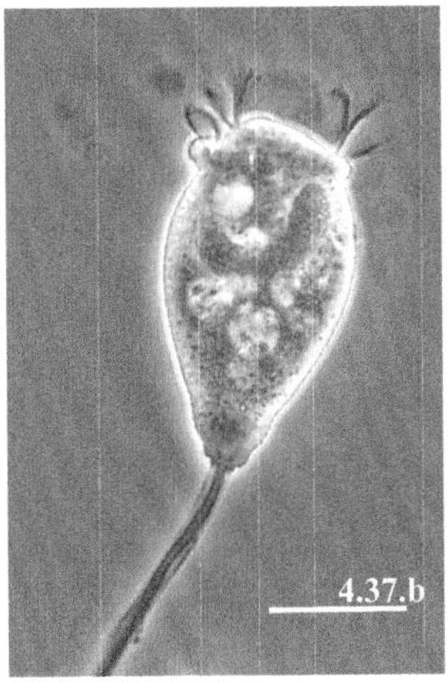

Figures 4.37.a and 4.37.b. ***Vorticella infusionum* complex.**

Vorticella microstoma complex

Size: 55 x 35 μm. Small zooid with 50-70 transverse striations (silver lines); linear or C-shaped macronucleus longitudinally placed; peristomial overture slightly smaller than the zooid maximum width. Ciliate associated to flocs.

V. microstoma was observed to be dominant in bad quality activated sludge, high wastewater flow or low organic load; it indicates a lack of dissolved oxygen in the aeration tank and a low effluent quality (Esteban *et al.* 1990, 1991; Madoni 1988, 1994a; Salvadó *et al.* 1995).

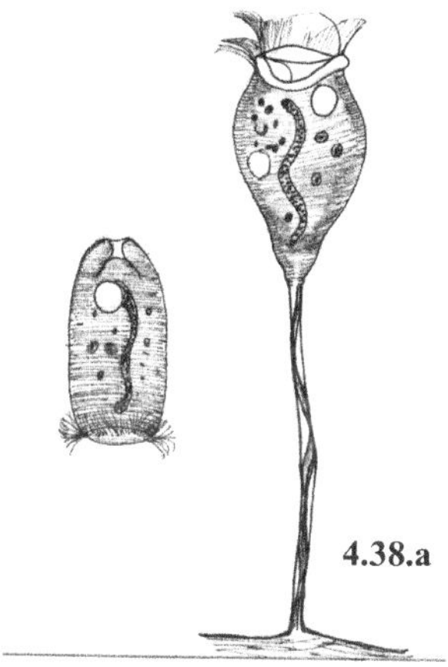

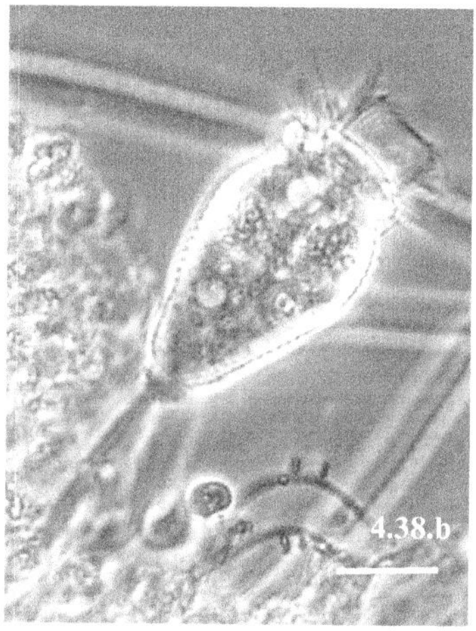

Figures 4.38.a and 4.38.b. ***Vorticella microstoma* complex.**

Vorticella convallaria complex

Size: 70 x 40 µm. Bell-like zooid with more than 100 transverse striations (silver lines); J- shaped macronucleus; the peristome has approximately the same width as the body cell. Sessile forms forming pseudocolonies are often observed on flocs or biofilms.

V. convallaria populations have been correlated with good performance of activated sludge plants, related also to high ammoniacal-N concentrations in both influent and effluent indicating a lack of nitrification in the process (Martín-Cereceda *et al.* 1996). It is usually associated to *Aspidisca cicada* (Madoni 1988). In RBC plants it is considered as a good bioindicator of organic matter removal through the system (Martín-Cereceda *et al.* 2002). *Vorticella convallaria*-complex has been also described in the third section of a wetland treating primary effluent plant and in the whole wetland treating secondary effluent, when the organic matter concentration was low (Puigagut *et al.* 2007).

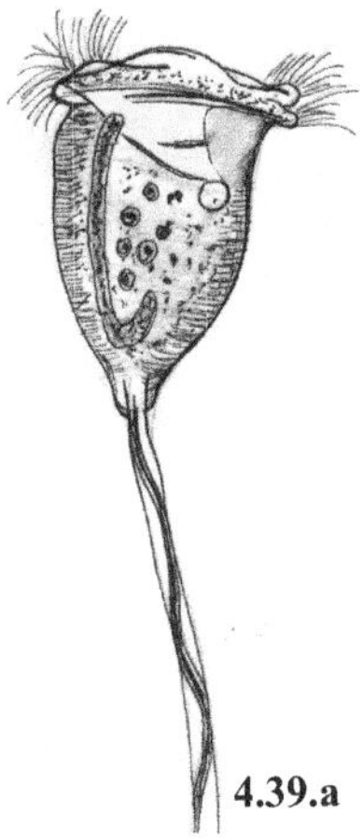
4.39.a

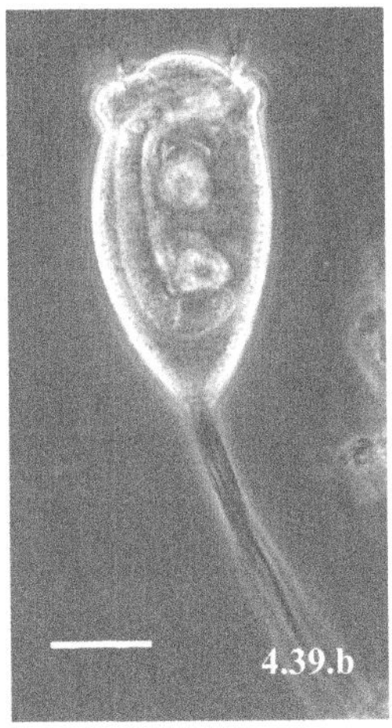

Figures 4.39.a and 4.39.b. ***Vorticella convallaria* complex**.

Vorticella campanula complex

Size: 105 x 65 μm. Bell-like zooid with a high number of transverse silver lines; large peristomial area extending outwards de cell; J- shaped macronucleus; numerous refractile cytoplasmic granules. Sessile forms rarely observed in WWTP.

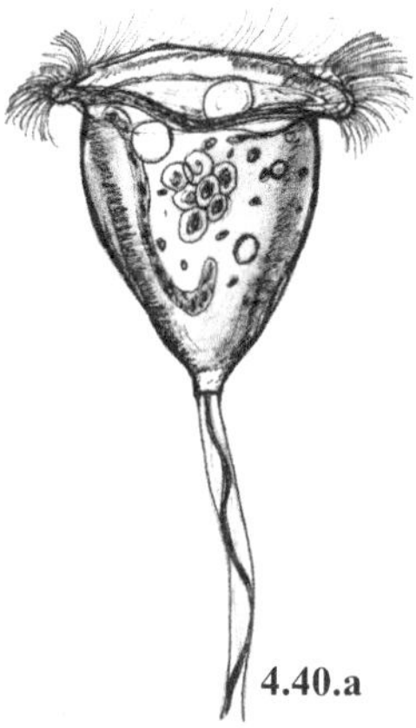

4.40.a

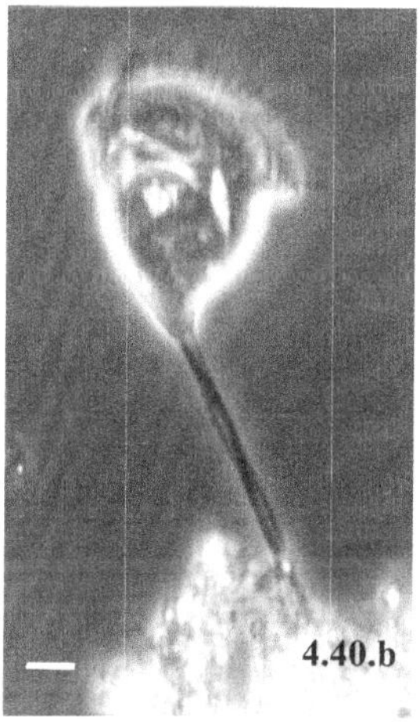

4.40.b

Figures 4.40.a and 4.40.b. ***Vorticella campanula* complex**.

Loricated peritrichs with one or two long zooids attached to the lorica without or with a short stalk. Lorica without valve: ***Vaginicola***

Species: ***Vaginicola cristallina*** Fromentel, 1874

Size: 200 x 100 µm. Transparent lorica with one or two cells; long zooids attached to the lorica by the posterior end (sometimes a short stalk is described); lengthened macronucleus. Ciliate attached to surfaces by the lorica.

Scarce in biological treatment reactors, it is associated with low loading or diluted influent (Madoni 1988).

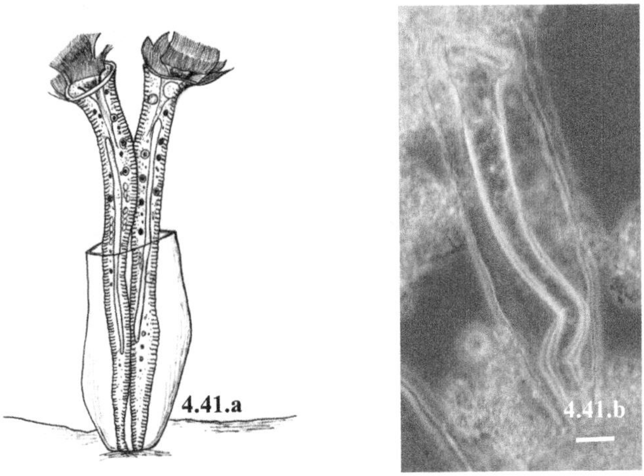

Figures 4.41.a and 4.41.b. ***Vaginicola cristallina*** Fromentel, 1874.

Lorica without stalk. One or two long zooids with a short stalk attaching to the lorica. Lorica with an anterior valve: ***Thuricola***

Species: ***Thuricola kellicottiana*** Stokes, 1887

Size: 330 x 30 µm. Cells attached to a lorica by a thin stalk (10 µm or more); membranous valve close the anterior end of the lorica when the zooids are retracted; visible transversal striation. This species is often confused *in vivo* with *T. folliculata* (with endosymbiotic algae and stalk shorter than 10 µm).

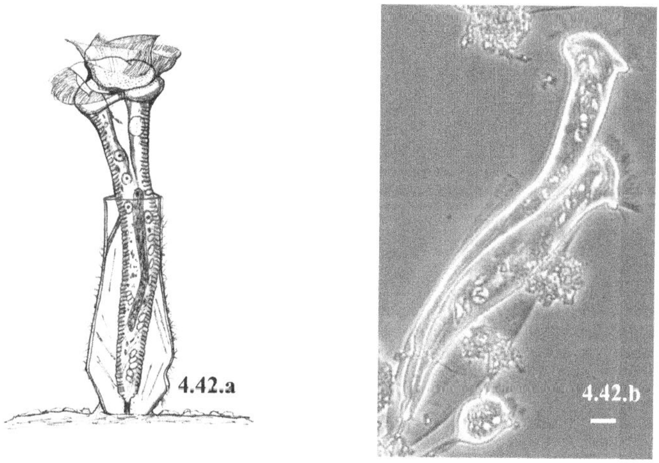

Figures 4.42.a and 4.42.b. ***Thuricola kellicottiana*** Stokes, 1887.

Colonial species with branched stalk with self-contained discontinuous spasmonemes (each stalked member of the colony contracts independently): ***Carchesium***

Species: ***Carchesium polypinum*** (Linnaeus, 1758) Ehrenberg, 1831

Zooid size: 120-80 μm. Colonies with a high number of bell-like zooids (they can reach a maximum width of 1mm); zooids are often bent; C-shaped macronucleus; transversal striation is no evident. Sessile colonies attached to the floc.

Appears abundantly in WWTP with low turbulence (Buck 1968; Pérez-Uz *et al.* 1998).

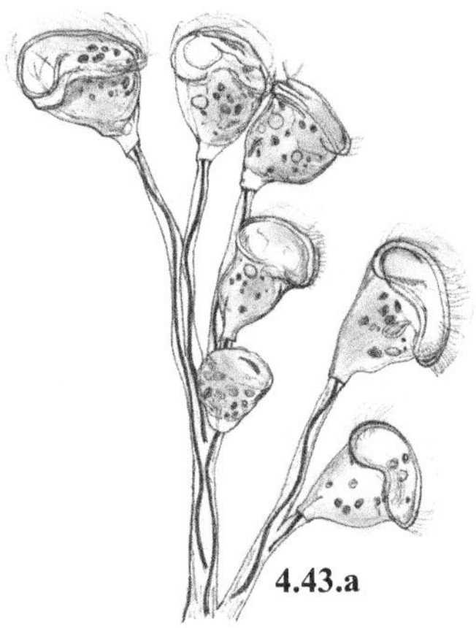

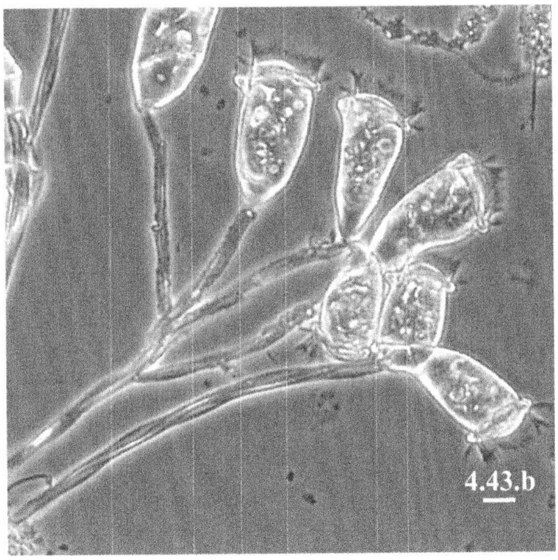

Figures 4.43.a and 4.43.b. ***Carchesium polypinum*** (Linnaeus, 1758) Ehrenberg, 1831.

Colonial species with branched stalk with continuous spasmonemes (all the zooids of the colony contract simultaneously): ***Zoothamnium***

Species: ***Zoothamnium procerius*** Kahl, 1935

Zooid size: 70 x 35 μm. Thick stalk with dicotomic branching (some specimens with conspicuous sticked detritus); prominent peristomial disc; transversally or obliquely placed macronucleus; the contractile vacuole is placed at the anterior pole, above the macronucleus. Stalked to the biofilms or flocs, the colonies present a variable number of zooids.

More frequent in rotating biological contactors (Pérez-Uz *et al.* 1998) than in activated sludge plants. Its occurrence in RBC systems is related to the adequate performance of the biological treatment, being also very sensitive to small changes in the inorganic parameters (Martín-Cereceda *et al.* 2002). Less common in activated sludge, however Madoni (1988) pointed out that the species of *Zoothamnium* are related with good depuration process and high quality effluent.

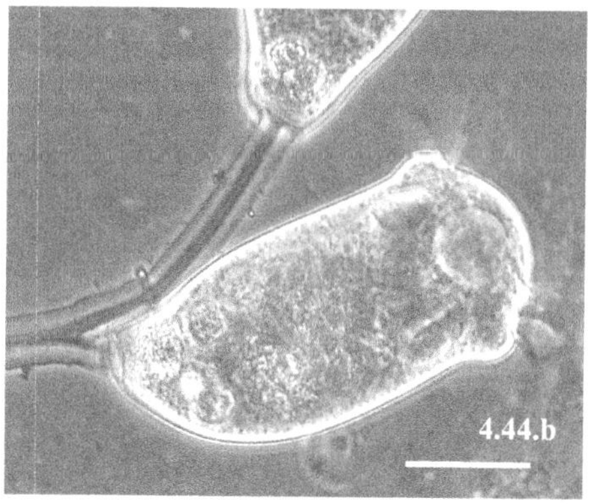

Figures 4.44.a and 4.44.b. ***Zoothamnium procerius*** Kahl, 1935.

Stalked peritrichs without spasmoneme (no contractile colonies). Colonial species, occasionally single cells: ***Epistylis***

Species: ***Epistylis plicatilis*** Ehrenberg, 1838

Size: 125 x 37 µm. Trumpet-like zooids with a prominent peristomial lip; transversal to longitudinal elongated macronucleus; longitudinal striation of the stalk slightly patent; characteristic "folds" at the posterior end in contracted cells; larvae with a "flying saucer" shape.

A massive presence in small colonies is associated with highly concentrated sludge (Madoni 1988).

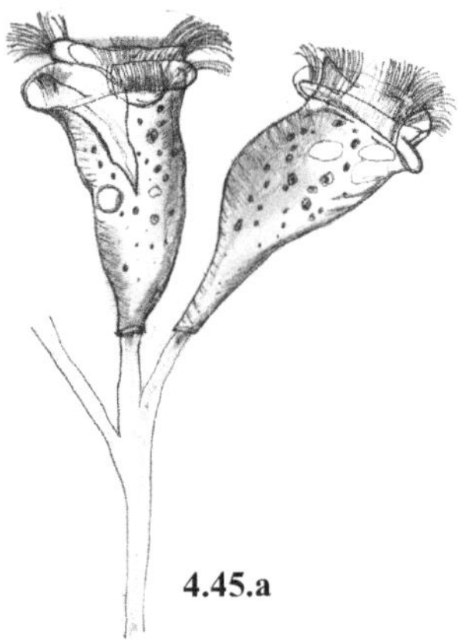

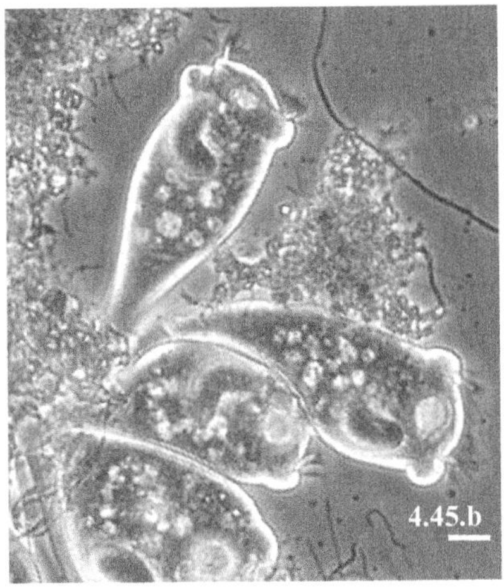

Figures 4.45.a and 4.45.b. *Epistylis plicatilis* Ehrenberg, 1838.

Species: ***Epistylis entzii*** Stiller, 1935

Size: 150 x 80 μm. Zooids with a narrow peristome and low number of transversal lines; prominent peristomial lip and peristomial disc; longitudinally striated stalk; transversal C-shaped long macronucleus surrounding the infundibulum. Floc associated colonies with few zooids.

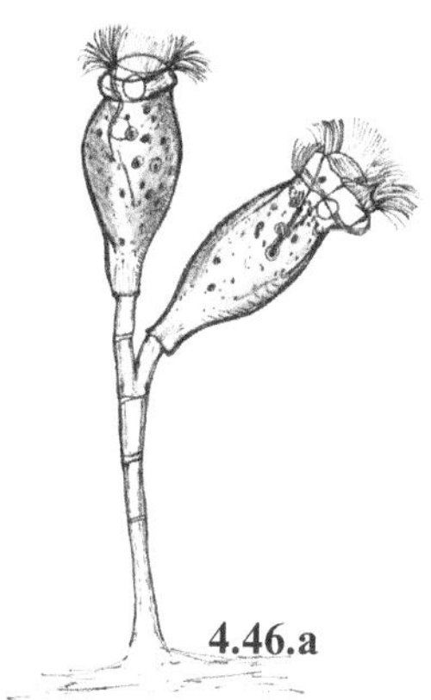

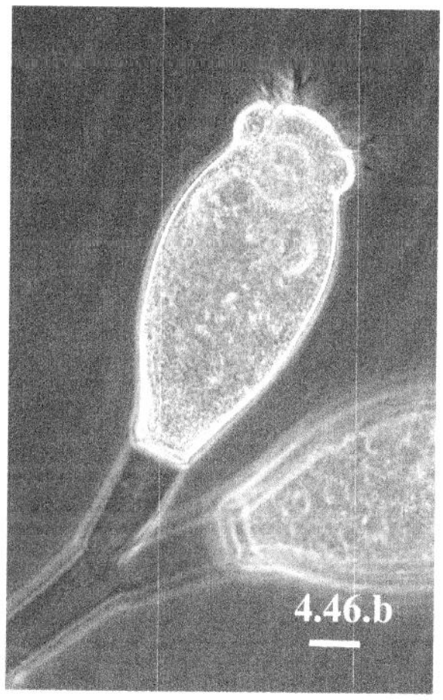

Figures 4.46.a and 4.46.b. ***Epistylis entzii*** Stiller, 1935.

Species: ***Epistylis chrysemidis*** Bishop and Jahn, 1941

Size: 160 x 85 µm. Zooids with prominent peristomial lip and an emerging peristomial disc; long infundibulum; transversal C-shaped macronucleus; stalks with fine longitudinal striation and transversal stretchmarks; anterior contractile vacuole. Sessile colonies attached to the flocs.

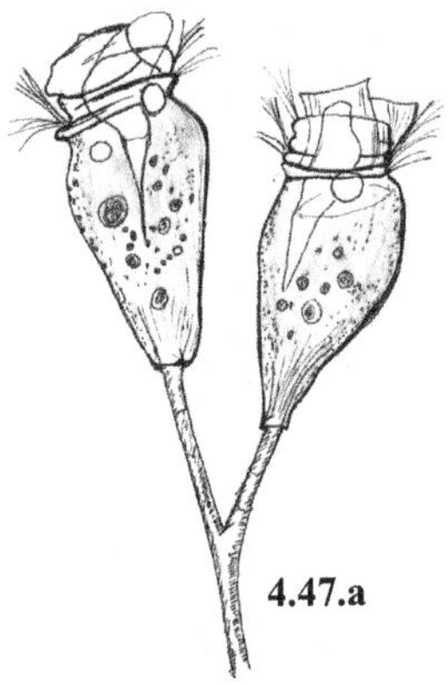

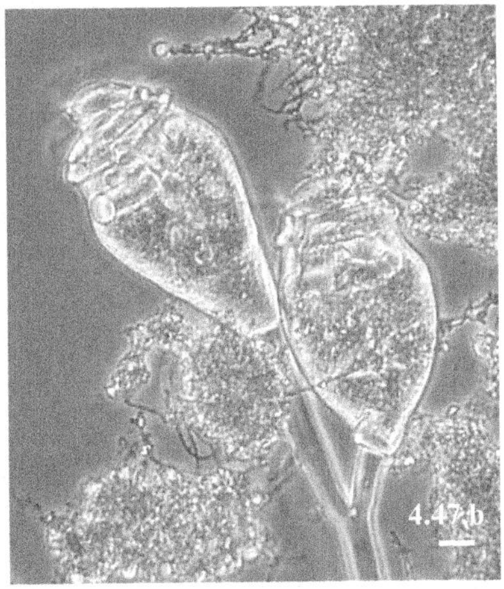

Figures 4.47.a, and 4.47.b). ***Epistylis chrysemidis*** Bishop and Jahn, 1941.

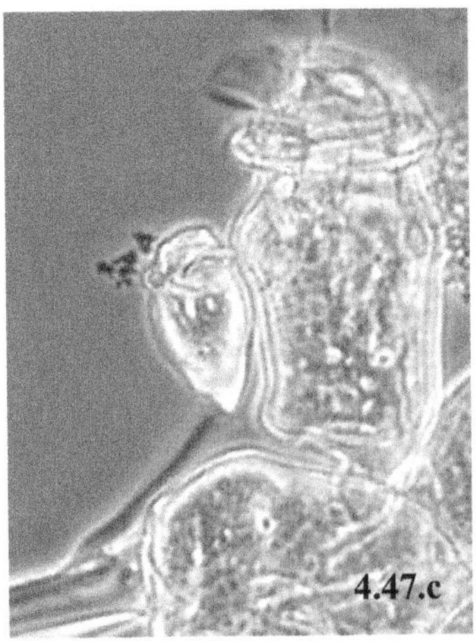

Foissner (1992) considers ***Epistylis balatonica*** as a synonymous species of *E. chrysemidis*. Although both species are very similar, *E. chrysemidis* is smaller (95 µm in length) and epibiotic over the stalk of other peritrichs (Figure 4.47.c).

Colonial species with no contractile stalks. Characteristic opercule emerging from the peristomial opening. Without a conspicuous peristomial lip: *Opercularia*

Species: *Opercularia coarctata* (Claparede and Lachmann, 1858) Roux, 1901

Size: 50 x 30 µm. Small colonies with 3-6 cells; buccal area with one emerging "opercule" and two spiral levels of peristomial ciliature; transversal macronucleus under the infundibulum; stalk smooth without longitudinal or transversal striations. Free-swimming long tubular telotroch larvae.

It has high tolerance to heavy metals (Madoni *et al.* 1996), found in sludges of plants treating industrial waste containing toxic substances (Esteban *et al.* 1990; Becares 1991); associated with high BOD effluent (Curds 1975) and ammoniacal-N concentration (Poole 1984; Madoni *et al.* 1993) and low effluent quality (Salvadó *et al.* 1995).

4.48.a

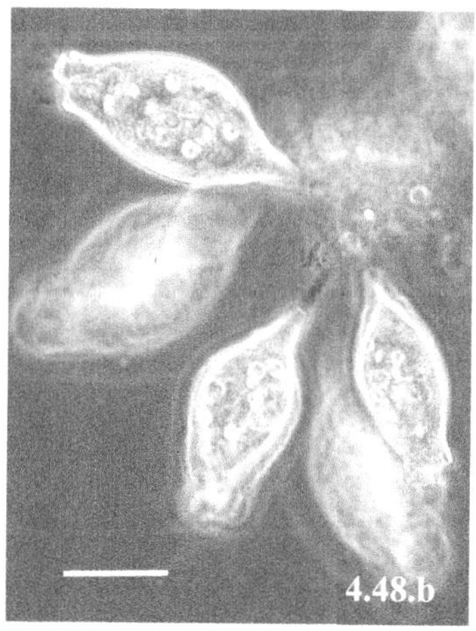

Figures 4.48.a and 4.48.b. ***Opercularia coarctata*** (Claparede and Lachmann, 1858) Roux, 1901.

Species: ***Opercularia microdiscum*** Fauré-Frèmiet, 1904

Size: 80 μm in length. Cylindrical zooids with a small opercule (its slightly protrudes from the peristomial overture); macronucleus surrounding the infundibulum; one contractile vacuole near the infundibulum in the anterior middle part of the cell; stalk with longitudinal striation.

This species tolerates low oxygen levels, being abundant at deficient aeration of biological reactors (Madoni 1988) and with high values of dissolved solids, low hydraulic retention time and low levels of oxygen, it could be considered as a good indicator of the quality of activated sludge (Esteban *et al.* 1990, 1991).

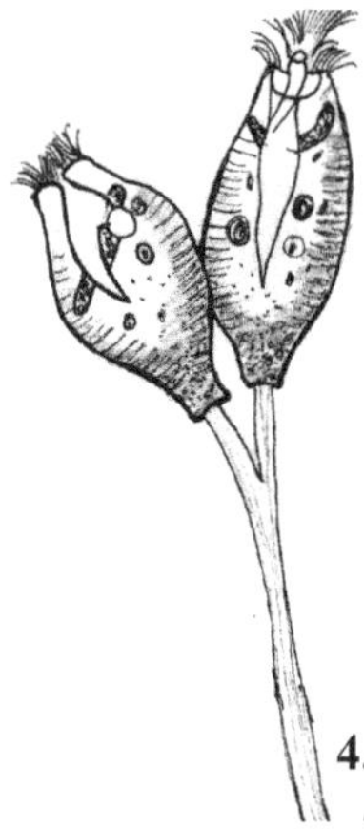

4.49.a

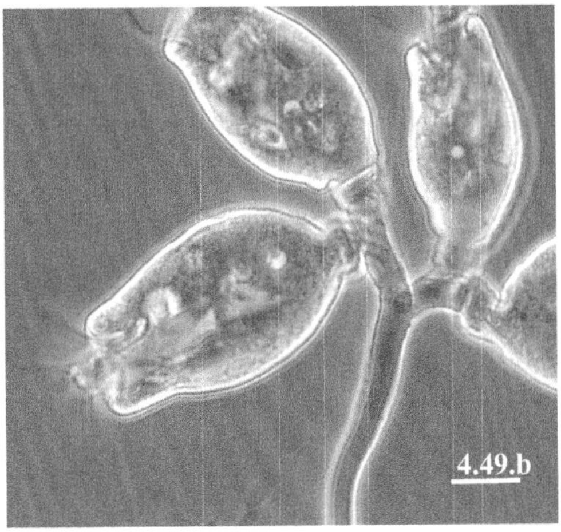

Figures 4.49.a and 4.49.b. ***Opercularia microdiscum*** Fauré-Frèmiet, 1904.

Species: ***Opercularia articulata*** Goldfuss, 1820

Size: 110 x 50 µm. Fusiform zooids with a prominent opercule; three spiral levels of peristomial ciliature; stalk with longitudinal and transversal striation; macronucleus surrounding infundibulum. Arborescent colonies with a high number of cells attached to bioaggregates.

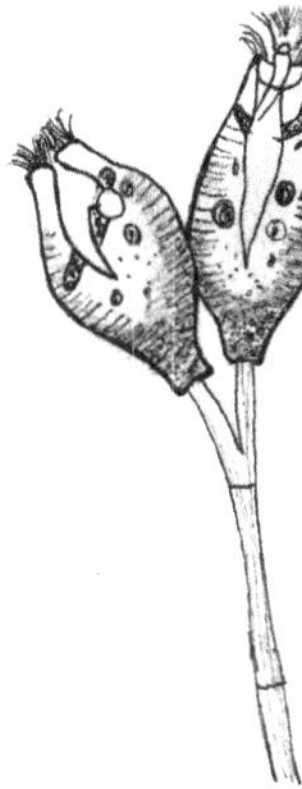

4.50.a

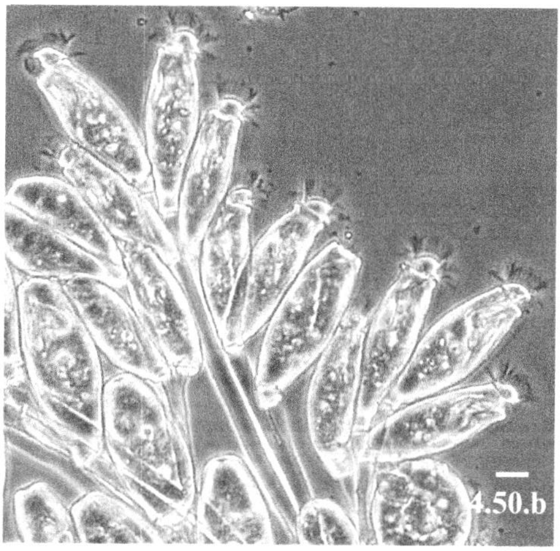

Figures 4.50.a and 4.50.b. ***Opercularia articulata*** Goldfuss, 1820.

Species: ***Opercularia curvicaula*** (Penard, 1922) Curds, 1964

Size: 60 μm in length. Colony has lengthened zooids and a prominent opercule; stalks have longitudinal and transversal striations; small transversal macronucleus and lemmon-shaped digestive vacuoles. Arborescent colonies with a high number of cells set on the flocs.

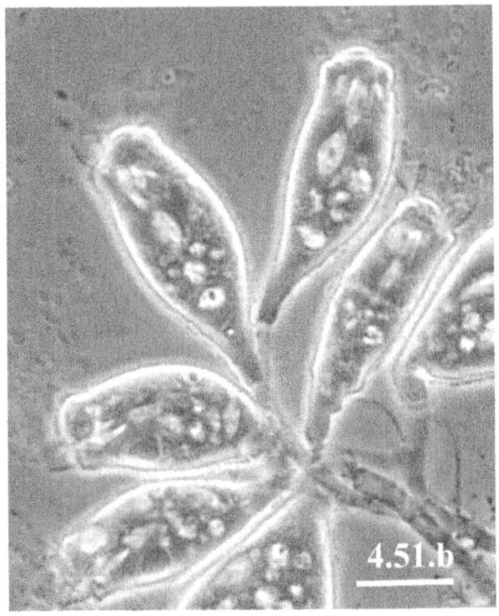

Figures 4.51.a and 4.51.b. ***Opercularia curvicaula*** (Penard, 1922) Curds, 1964.

4.9 PRORODONTIDS

Swimming ciliates. Rarely abundant in WWTP. Staining of cells is recommended for species identification.

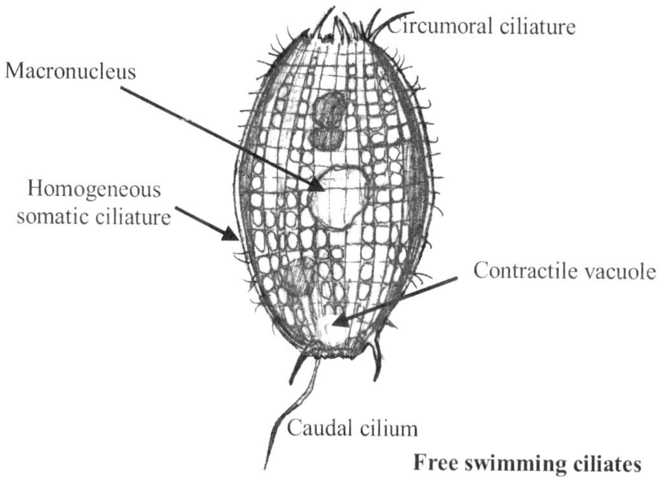

Figure 4.52. General characteristics of prorodontids.

Apical cytostome. Homogeneous somatic ciliature. One or more caudal cilia. Alveolar plates regularly disposed on the ciliate body. Posterior contractile vacuole: ***Coleps***

Species: ***Coleps hirtus*** (Müller, 1786) Nitzsch, 1827

Size: 50 x 25 µm. Oval dark cells with three posterior spines; circumoral ciliature surrounding the apical cytostome; longer cilia compared to somatic ciliature and one long caudal cilium. Fast turning around movements when swimming freely, it can be observed clearly when it is eating over surfaces or detritus.

Species with low tolerance to ammoniacal nitrogen, only reaches moderate abundance when nitrifying process is effective (Bick 1972; Poole 1984; Madoni *et al.* 1993; Martín-Cereceda *et al.* 1996).

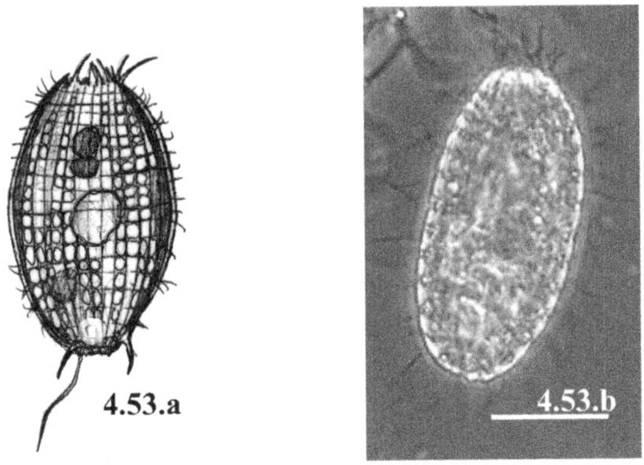

Figures 4.53.a and 4.53.b. ***Coleps hirtus*** (Müller, 1786) Nitzsch, 1827.

4.10 HAPTORIDS

Swimming ciliates. Rarely with high abundance in WWTP. Staining of cells is recommended for species identification.

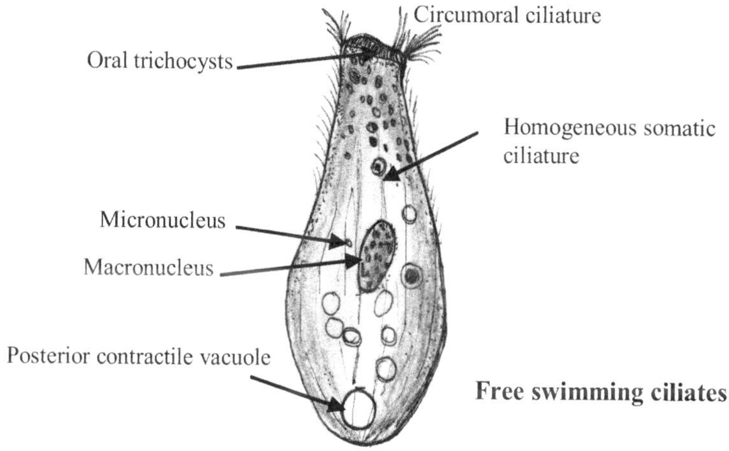

Figure 4.54. General characteristics of haptorids.

Apical or subapical oval cytostome with a short cytopharyngeal basket within the anterior pole. Homogeneous somatic ciliature. Posterior contractile vacuole: ***Enchelys***

Species: ***Enchelys gasterosteus*** Kahl, 1926

Size: 65 x 30 µm. Ovoid cell with a flat anterior end slightly emerging from the cell in which the cytostome is located; large ovoid macronucleus; posterior contractile vacuole. Free-swimming cells, occasionally feed on the floc or biofilms surfaces.

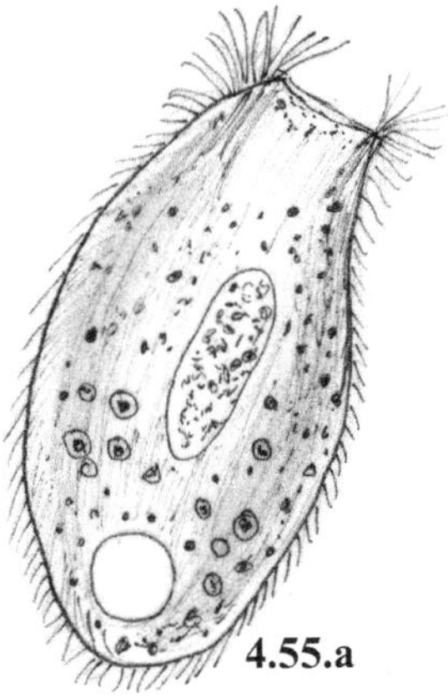

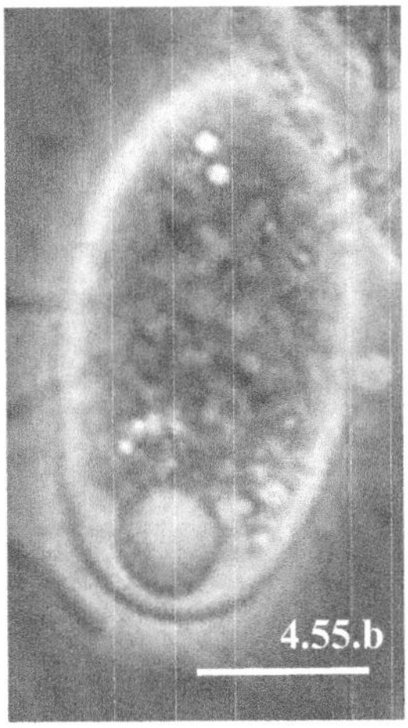

Figures 4.55.a and 4.55.b. ***Enchelys gasterosteus*** Kahl, 1926.

Long retractile anterior neck-like extension (proboscis). Apical circumoral ciliature surrounding the cytostome. Homogeneous somatic ciliature: ***Lacrymaria***

Species: ***Lacrymaria olor*** (Müller, 1786) Bory, 1824

Size: 400 x 25 µm (up to 1 mm extended). Oblique, slightly spiral somatic kineties; cytostome at the end of the retractile proboscis; two spherical macronuclei; posterior contractile vacuole. Swimming ciliate, when it is feeding the movement is slower and the proboscis bends continuously.

4.56.a

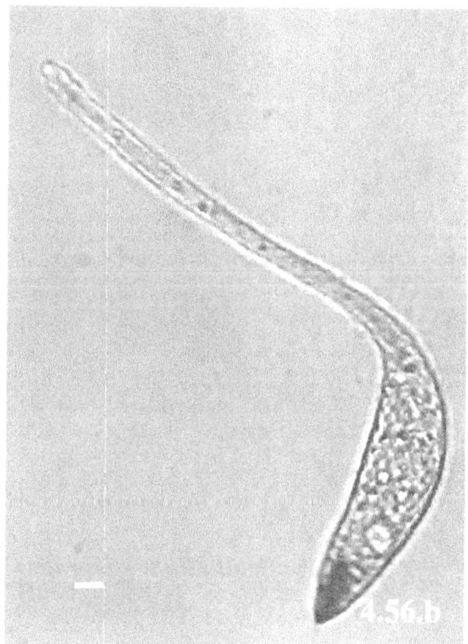

Figures 4.56.a and 4.56.b. ***Lacrymaria olor*** (Müller, 1786) Bory, 1824.

Extensible latero-apical cytostome with patent fibrillar toxicysts. Homogeneous somatic ciliature: ***Spathidium***

Species: ***Spathidium spathula*** (Müller, 1786) Woodruff and Spenser, 1922

Size: 90 x 40 µm. Bag-like cells with a lengthened anterior cytostome (it can extend to feed on larger organisms); ovoid to lengthened macronucleus; posterior contractile vacuole, homogenous patent somatic ciliature. Predatory ciliate.

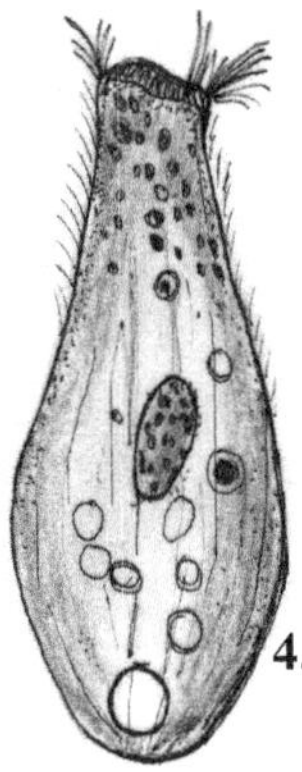

4.57.a

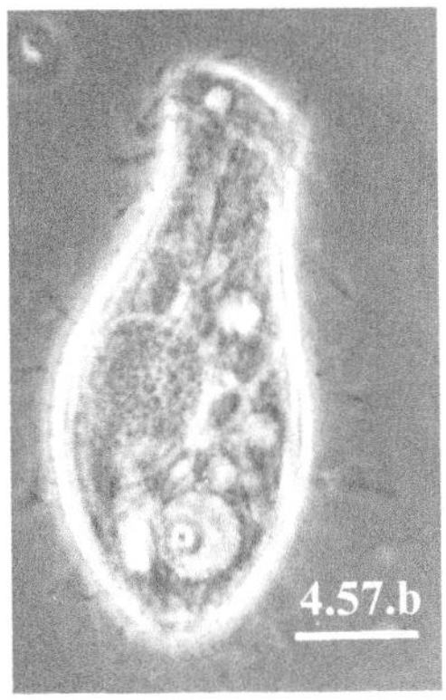

Figures 4.57.a and 4.57.b. ***Spathidium spathula*** (Müller, 1786) Woodruff and Spenser, 1922.

Cytostome at the basis of an anterior proboscis. Patent cytopharingeal basket. Homogeneous somatic ciliature: ***Dileptus***

Species: ***Dileptus anguillula*** Kahl, 1931

Size: 140 x 20 µm. Fusiform cells with a small proboscis; several dorsal contractile vacuoles; two macronuclear nodes with a variable shape (lengthened, ovoid, moniliform). Species rare in WWTP.

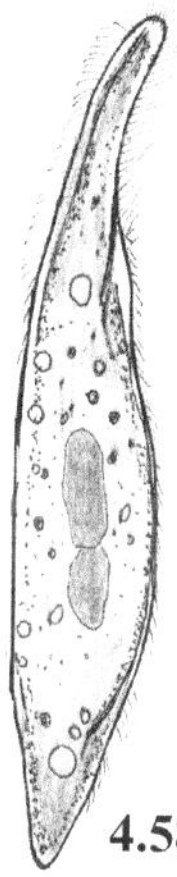

4.58.a

Figures 4.58.a and 4.58.b. ***Dileptus anguillula*** Kahl, 1931.

4.11 PLEUROSTOMATIDS

Species of this group are common in biological reactors, both free-swimming and floc-associated ciliates. Ciliary pattern is necessary to determine genera and species, so staining procedures are recommended.

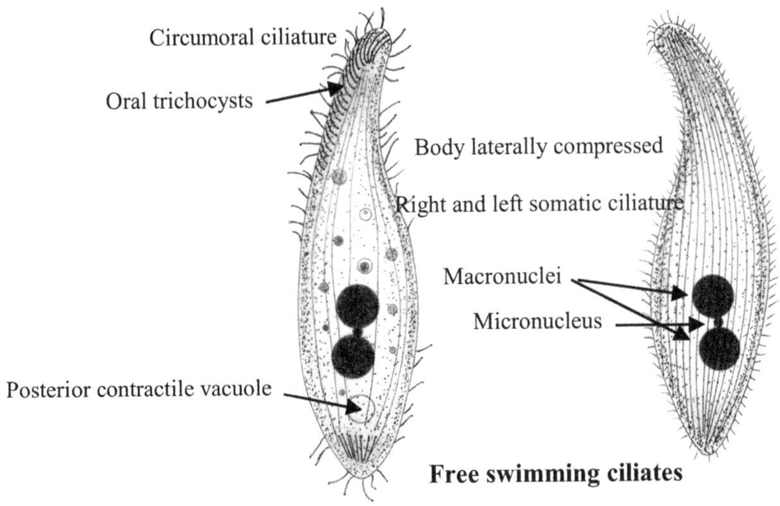

Figure 4.59. General characteristics of pleurostomids.

Lanceolated cells with a lateral body compression. Different somatic ciliature arrangement on right and left sides of the cell. Lateral cytostome with perioral ciliature (not curved). Two macronuclei and one central micronucleus usually visible with optical microscopy: ***Litonotus***

Species: ***Litonotus lamella*** (Ehrenberg) Schewiakoff, 1896

Size: 75 x 18 µm. Lanceolated cells with an antero-lateral slit-like cytostome; toxicysts at the oral zone and randomly distributed around the cell; numerous somatic kineties differently distributed at both cell sides; posterior contractile vacuole; two macronuclei and a large micronucleus placed between the macronuclear nodes. Free-swimming occasionally associated to the floc surface.

This species appears associated to a poor settling sludge (Martín-Cereceda *et al.* 1996; Lee *et al.* 2004). Its occurrence in RBC plants is related to efficient organic matter removal through the system (Martín-Cereceda *et al.* 2002).

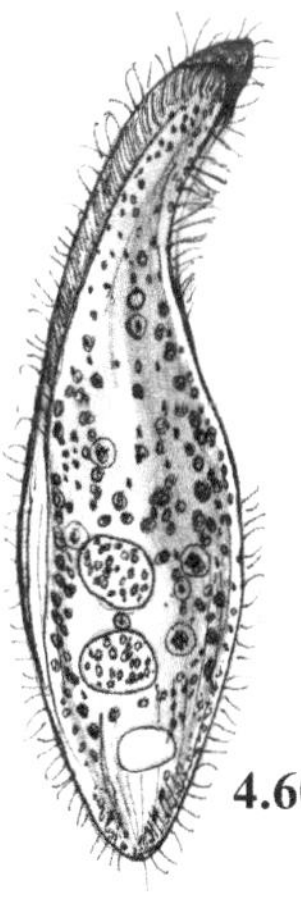

4.60.a

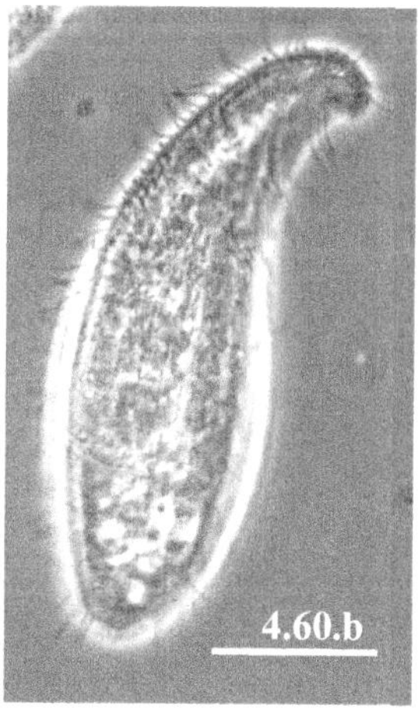

Figures 4.60.a and 4.60.b. ***Litonotus lamella*** (Ehrenberg) Schewiakoff, 1896.

Lanceolated ciliates with a characteristic torsion at the anterior end. Lateral cytostome and perioral ciliature (curved at the anterior part). Few somatic kineties. Two macronuclei and one central micronucleus: *Acineria*

Species: ***Acineria uncinata*** Tucolesco, 1962

Size: 30 x 14 µm. Small cells with an anterior torsion to the left side; short lateral cytostome; long conspicuous perioral cilia at the anterior torsion; posterior contractile vacuole. It has been confused occasionally with *Trachelophyllum pusilum* as has been noted by Augustin and Foissner (1992a) and Martín-Cereceda *et al* (1995).

In activated sludge plants *A. uncinata* has been found to be negatively related to BOD_5 and COD_5 loads in the inflow water (Martín-Cereceda *et al.* 1995). Its occurrence in RBC systems is related to a decrease of the organic factor, so that seems to be a good indicator of the biological treatment efficiency (Martín-Cereceda *et al.* 2002).

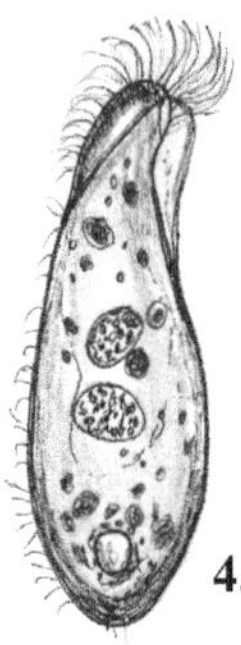

4.61.a

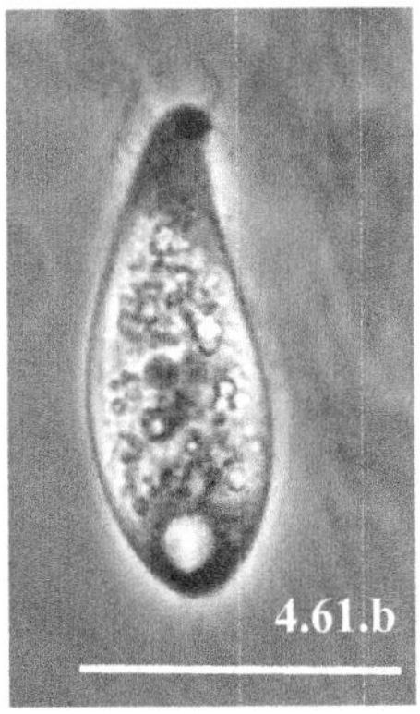

Figures 4.61.a and 4.61.b. ***Acineria uncinata*** Tucolesco, 1962.

4.12 PHYLOFARINGIDS

Species of this group are common in biological treatment plants. Ciliary pattern is necessary to determine genera and species, so staining procedures are recommended.

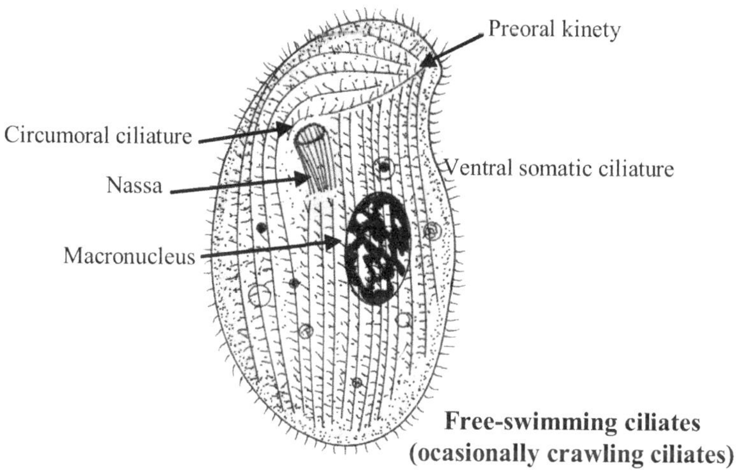

Figure 4.62. General characteristics of phylofaringids.

Dorso-ventrally flattened body. Ventral somatic ciliature at the right and left sides of the oral area (the anterior right somatic kineties curves preorally). Continuous preoral kinety and two circumoral kineties: *Chilodonella*

Species: ***Chilodonella uncinata*** (Ehrenberg, 1838) Strand, 1928

Size: 45 x 28 µm. Flattened body with a posterior part more swollen than the anterior one (it seems to be clearer at the microscope); comma-like nassa; oval macronucleus; two contractile vacuoles (anterior and posterior) in opposed margins of the cell; short anterior and oblique kinety in the dorsal side of the cell. Swimmer ciliate, occasionally is moving over bioaggregates on its ventral side.

It could be a useful species for indication of good effluent quality (Lee *et al.* 2004). Species associated to nitrifying conditions in activated sludge plants (Madoni *et al.* 1993) and with a very high sensitivity to heavy metals (Madoni *et al.* 1996).

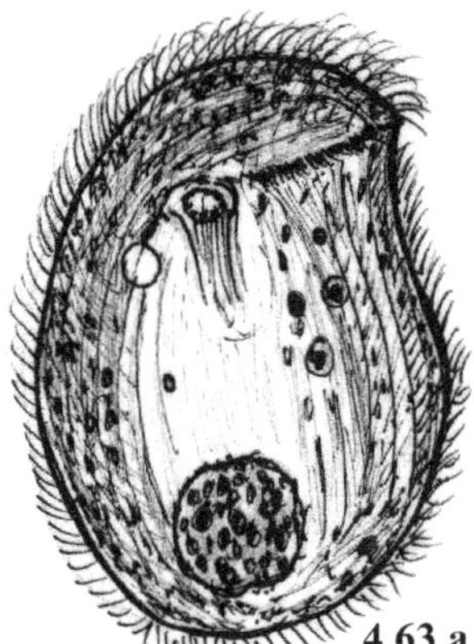

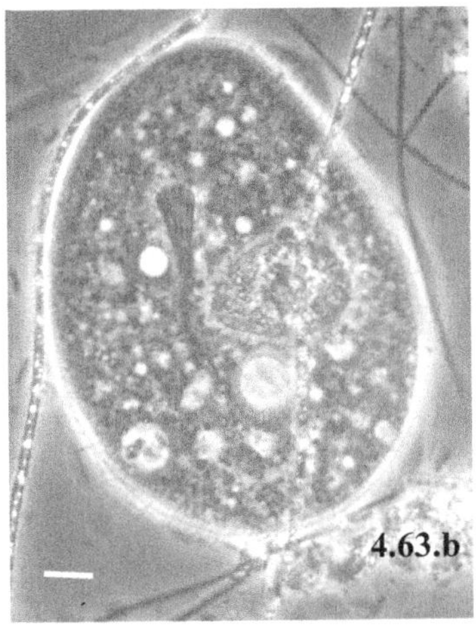

Figures 4.63.a and 4.63.b. ***Chilodonella uncinata*** (Ehrenberg, 1838) Strand, 1928.

Dorso-ventrally flattened body. Somatic ciliature at the right and left sides of the ventral side. Fragmented preoral kinety with patent cilia: ***Pseudochilodonopsis***

Species: ***Pseudochilodonopsis fluviatilis*** Foissner, 1988

Size: 60 x 30 μm. Dorso-ventrally flattened body with ventral somatic kineties at right and left of the oral area; fragmented preoral kinety; two contractile vacuoles anterior and posterior in opposite sides of the cell; the nassa is straight. It moves around the floc or freely in the mixed liquor.

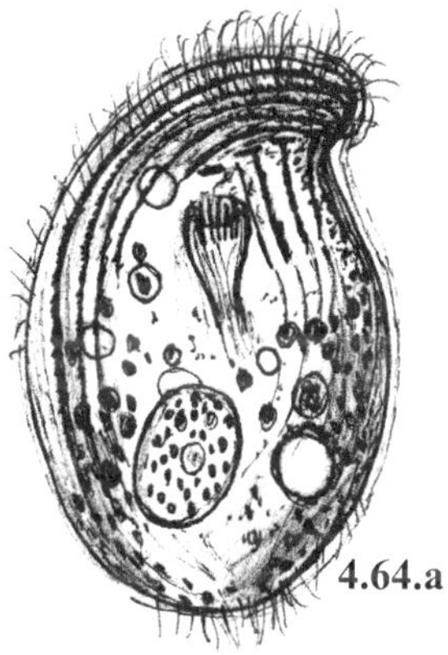

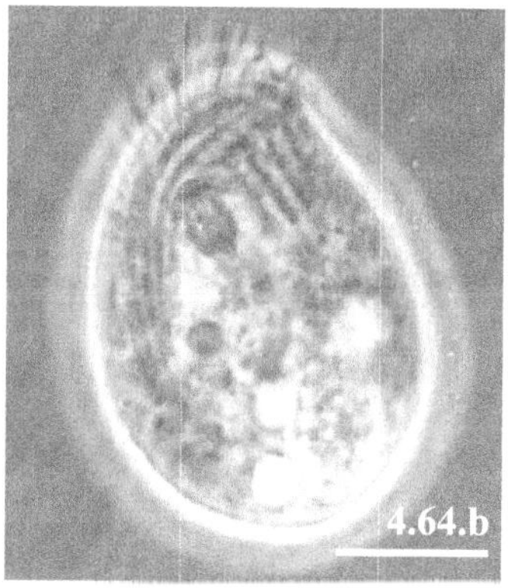

Figures 4.64.a and 4.64.b. ***Pseudochilodonopsis fluviatilis*** Foissner, 1988.

Dorso-ventrally flattened body. Homogeneous ventral somatic ciliature. Continuous preoral kinety and two circumoral kineties: ***Trithigmostoma***

Species: ***Trithigmostoma cucullulus*** (Müller, 1786) Jankowski, 1967

Size: 90 x 35 μm. Large cells with a straight prominent nassa; the dorsal surface has a very small anterior and oblique kinety; several contractile vacuoles; ovoid macronucleus. It moves slowly in the mixed liquor or on the surfaces.

This species tolerates a wide range of environmental conditions (Bick 1972).

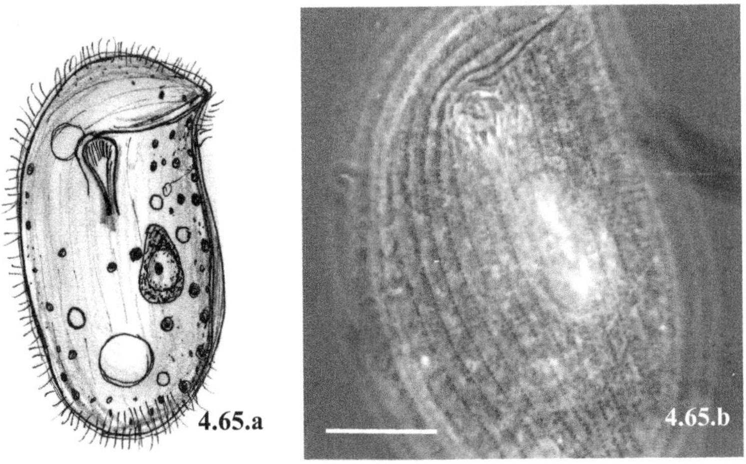

Figures 4.65.a and 4.65.b. ***Trithigmostoma cucullulus*** (Müller, 1786) Jankowski, 1967.

Dorso-ventrally flattened cell with an attaching posterior spine-like appendix. Somatic ciliature at left side of the ventral side. Two circumoral and two preoral kineties: ***Trochilia***

Species: ***Trochilia minuta*** (Roux, 1901) Kahl, 1931

Size: 25 x 14 μm. Small cells with a long straight cytopharingeal nassa; two contractile vacuoles (anterior and posterior) located both on the left side of the cell; long anterior cilia. Feeds on filamentous bacteria, being generally associated to the floc.

Species associated to nitrifying conditions in activated sludge plants (Madoni *et al.* 1993) and with a very high sensitivity to heavy metals (Madoni *et al.* 1996).

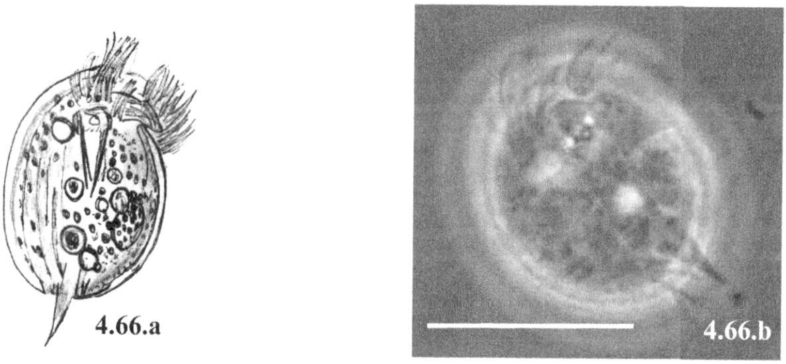

Figures 4.66.a and 4.66.b. ***Trochilia minuta*** (Roux, 1901) Kahl, 1931.

4.13 SUCTORIDS

Sessile vegetative cells without ciliature but with tentacles with toxicyst involved in predation of other protists. Cells immobile attached to the substrate, with a non-contractile stalk or without it. Species are common in WWTP but with low abundances.

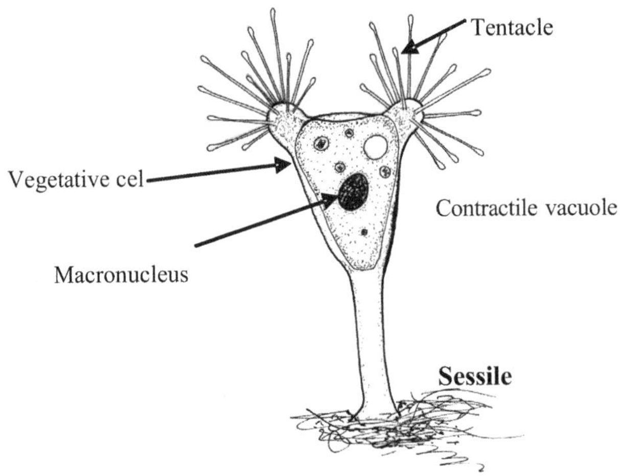

Figure 4.67. General characteristics of suctorids.

Spherical trophonts, usually without lorica. Tentacles homogenously distributed: ***Podophrya***

Species: ***Podophrya fixa*** (Müller, 1786) Ehrenberg, 1833

Size: 50-75 μm (diameter). Spherical cells with stalk (rarely without it); capitate tentacles randomly placed around the cell; oval macronucleus; single contractile vacuole.

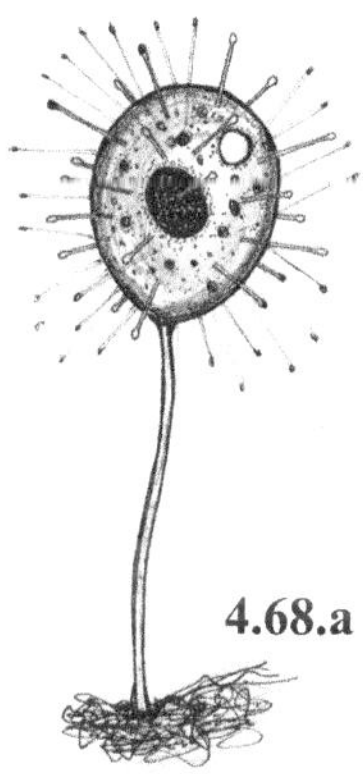

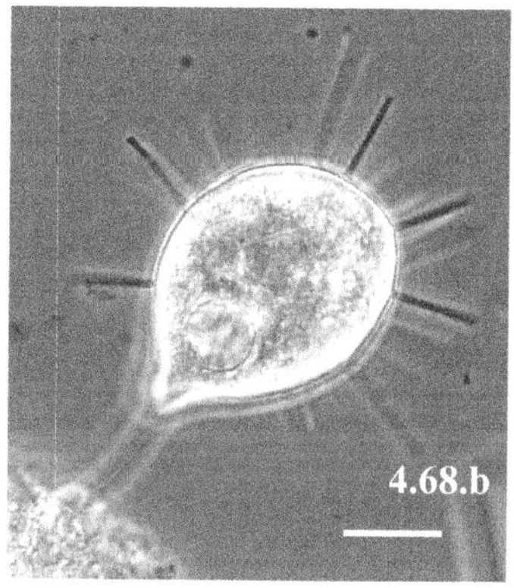

Figures 4.68.a and 4.68.b. ***Podophrya fixa*** (Müller, 1786) Ehrenberg, 1833

Sessile ciliates without lorica. Fascicles of fine tentacles: ***Tokophrya***

Species: ***Tokophrya lemnarum*** (Stein, 1859) Entz, 1903

Size: 70 μm in length. Triangular-shaped cells with two anterior clusters of opposed capitated tentacles; stalks variable in length; one or two anterior contractile vacuoles; ovoid macronucleus.

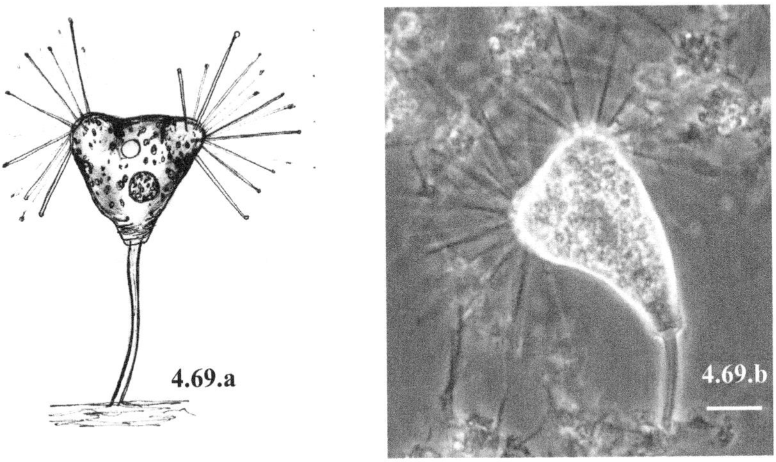

Figures 4.69.a and 4.69.b. ***Tokophrya lemnarum*** (Stein, 1859) Entz, 1903.

Sessile ciliates with a lorica. Two groups of tentacles placed on two anterior prominences: ***Acineta***

Species: ***Acineta tuberosa*** (Pallas, 1766) Ehrenberg, 1833

Size: 70 μm in length. Triangular-shaped cells with clusters of tentacles in two antero-lateral lobular prominences; loricated species with a stalk; macronucleus rounded and one anterior contractile vacuole. Synonym of *A. fluviatilis* Stokes, 1885.

This species is associated with a high quality effluent (Salvadó *et al.* 1995).

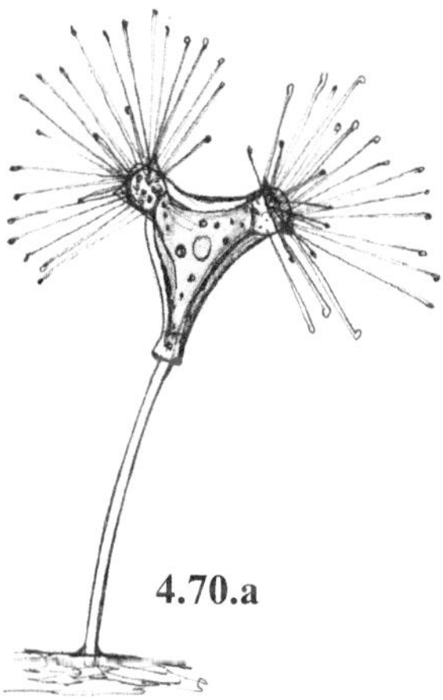

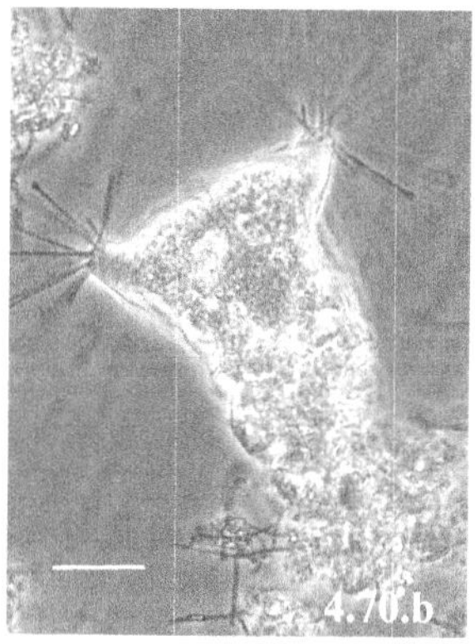

Figures 4.70.a and 4.70.b. ***Acineta tuberosa*** (Pallas, 1766) Ehrenberg, 1833.

Sessile loricated ciliates. Clusters of fine tentacles: ***Metacineta***

Species: ***Metacineta mystacina*** (Ehrenberg, 1831) Butschli, 1889

Size: 65 µm in diameter. Cells with six groups of long tentacles and a stalked lorica; one contractile vacuole and a spherical macronucleus.

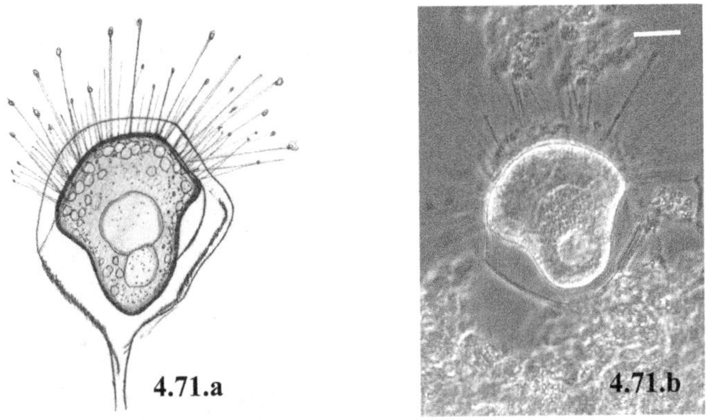

Figures 4.71.a and 4.71.b. ***Metacineta mystacina*** (Ehrenberg, 1831) Butschli, 1889.

Sessile ciliates with a stalk. Several groups of fine tentacles (lateral, anterior and posterior): ***Multifasciculatum***

Species: ***Multifasciculatum elegans*** Goodrich and Jahn, 1943

Size: 70 x 35 μm in diameter. Stalked ovoid cells with clusters of long tentacles at anterior and posterior ends plus two lateral groups; two lateral and opposed contractile vacuoles (over and under the equatorial zone); elongated macronucleus.

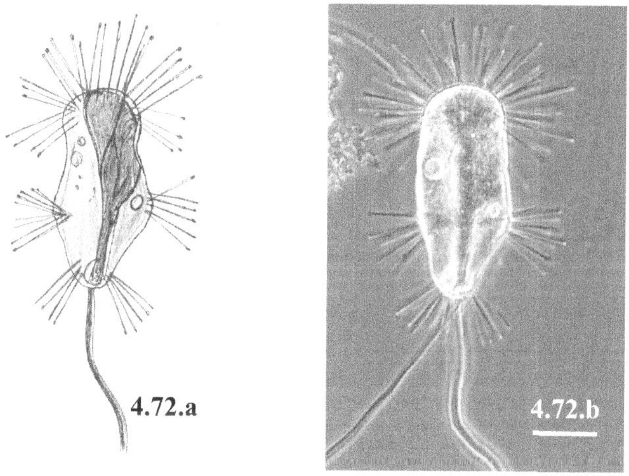

Figures 4.72.a and 4.72.b. ***Multifasciculatum elegans*** Goodrich and Jahn, 1943.

5

Bibliography

Adl, S.M., Simpson, A.G.B., Farmer, M.A., Andersen, R.A., Anderson, O.R., Barta, J.R., Browser, S.S., Brugerolle, G., Fensome, R.A., Fredericq, S., James, T.Y., Karpov, S., Kugrens, P., Krug, J., Lane, C.E., Lewis, L.A., Lodge, J., Lynn, D.H., Mann, D.G., McCourt, R.M., Mendoza, L., Moestrup, O., Mozley-Standridge, S.E., Nerad, T.A., Shearer, C.A., Smirnov, A.V., Spiegel, F.W. and Taylor, M.F.J.R. (2005) The new higher level classification of eukaryotes with emphasis on the taxonomy of protists. *J. Eukaryot. Microbiol.* **52**, 399-451.

Al-Shahwani, S.M. and Horan, N.J. (1991) The use of protozoa to indicate changes in the performance of activated sludge plants. *Water Res.* **25**, 633-638.

Anderson, O.R. (1988) Perspective on Ciliate Ecology. In *Comparative Protozoology, Ecology, Physiology, Life History*, pp. 109-130, Springer-Verlag, New York Inc.

Arregui, L., Muñoz-Fontela, C., Serrano, S., Barasoain, I. and Guinea, A. (2002) Direct visualization of the microtubular cytoskeleton of ciliated protozoa with a fluorescent taxoid. *J. Eukaryot. Microbiol.* **49**, 312-318.

Arregui, L., Muñoz-Fontela, C., Guinea, A. and Serrano, S. (2003) FLUTAX facilitates visualization of the ciliature of oxytrichid hypotrichs. *Eur. J. Protistol.* **39**, 169-172.

Arregui, L., Serrano, S., Linares, M., Pérez-Uz, B. and Guinea, A. (2007) Ciliate contributions to bioaggregation: laboratory assays with axenic cultures of *Tetrahymena thermophila*. *Int. Microbiol.* **10**, 91-96.

by Susana Serrano, Lucia Arregui, Blanca Perez-Uz, Pilar Calvo and Almudena Guinea.
ISBN: 9781843391715. Published by IWA Publishing, London, UK.

Augustin, H., Foissner, W. and Adam, H. (1984) An improved pyridinated silver carbonate method which needs few specimens and yields permanent slides of impregnated ciliates (Protozoa, Ciliophora). *Mikroskopie* **41**, 134-137.

Augustin, H. and Foissner, W. (1992a) Morphology and ecology of some ciliates (Protozoa, Ciliophora) from activated sludge. *Arch. Protistenkd.* **141**, 243-283.

Augustin, H. and Foissner, W. (1992b) Enumeration of protists and small metazoans in activated sludge. In *Protocols in Protozoology*. (ed. J.J. Lee and A.T. Soldo), pp. B-8.1, 8.2, Society of Protozoology, Kansas.

Becares, E. (1991) Microfauna of an activated sludge pilot plant treating effluents from a pharmaceutic industry. In *Biological Approach to Sewage: Current Status and Perspectives* (ed. P. Madoni), pp. 105-108, Perugia.

Berger, H. (1999) Monograph of the Oxytrichidae (Ciliophora, Hypotrichia). *Monographiae Biologicae* **78**, Kluwer Academic Publishers, The Netherlands.

Berger, H., Foissner, W. and Adam, H. (1984) Taxonomie, biometrie und morphogenese einiger terricoler Ciliaten (Protozoa, Ciliophora). *Zool. Jb. Syst.* **111,** 339-367.

Bernhard, D., Stechmann, A., Foissner W., Ammermann, D., Hehn M. and Schlegel, M. (2001) Phylogenetic relationships within the class Spirotrichea (Ciliophora) inferred from small subunit rRNA gene sequences. *Mol. Phylogenet. Evol.* **21**, 86-92.

Berri, A., Casaschi, R. (1991) Observation of the microfauna and evaluation of nitrification and denitrification activity of biofilm on the RBC. In *Biological Approach to Sewage: Current Status and Perspectives* (ed. P. Madoni), pp. 105-108, Perugia.

Bick, H. (1963) A review of central European methods for the biological estimation of water pollution rates. *B. World Health Organ.* **29**, 401-413.

Bick, H. (1972) *Ciliated protozoa.* World Health Organization, Geneva.

Bishop P.L. and Kinner N.E. (1986) Aerobic fixed-film processes. ***In:*** *Biotechnology*, 1st edn. (ed. Schönborn), vol. 8, pp. 113-176, VCH Verlagsgesellschaft, Weinheim.

Bitton, G. (2005) *Wastewater microbiology,* 3rd edn., Wiley-Liss Pub., New York.

Borror, A.C. and Hill, B.F. (1995) The Order Euplotida: taxonomy, with division of *Euplotes* into several genera. *J. Eukaryot. Microbiol.* 42, 457-466.

Buck, H. (1968) Die ciliaten des belebtschlammes inihrer abhängigkeit vom klärverfahren. ***In:*** *Abwasser-, Fischerei- und Flussbiologie,* 2nd edn., vol 5, Oldenbourg, Munich.

Cairns, J. Jr (1978) Zooperiphyton (especially protozoa) as indicators of water quality. *T. Am. Microsc. Soc.* **97**, 44-49.

Cairns, J. Jr and Pratt J.R. (1988) Gathering time-dependent information: An optimization problem. *T. Am. Microsc. Soc.* **107**, 1-11.

Cardinaletti, M. and Zitelli, A. (1991) Distribuzione tassonomica della microfauna in una unità da laboratorio a fanghi attivati ricevente liquami urbano-industriali addizionati di zinco solubile. Parte II. In *Biological Approach to Sewage: Current Status and Perspectives* (ed P. Madoni), pp. 59-70, Perugia.

Cavalier-Smith, T. (1998) A revised six-kingdom system of life. *Biol. Rev.* **73**, 203-266.

Cavalier-Smith, T. (2002) The phagotrophic origin of Eukaryotes and phylogenetic classification of Protozoa. *Int. J. Syst. Evol. Micr.* **52**, 297-354.

Chatton, E. and Lwoff, A. (1930) Imprégnation, par diffusion argentique, de l'infraciliature des ciliés marins et d'eau douce, après fixation cytologique et sans desiccation. *C. r. Séanc. Soc. Biol.* **104**, 834-836.

Chatton, E. and Lwoff, A. (1936) Techniques pour l'étude des protozoaires, spécialement de leurs structures superficielles (cinétome et argyrome). *B. Soc. Fr. Microsc.* **5**, 25-39.

Chung, J. and Strong, P. (1991) Microbiological study of ten New Jersey rotating biological contactor wastewater treatment plants. *Res. J. Water Pollut. C.* **63**, 35-43.

Corliss, J.O. (1953) Silver impregnation of ciliated protozoa by the Chatton-Lwoff technique. *Stain Technol.* **28**, 97-100

Corliss , J. O. (1979) *The ciliated protozoa: Characterization, classification and guide to the literature*, 2nd edn, Pergamon Press, Oxford.

Corliss, J.O. (1998) Classification of protozoa and protists: the current status. In *Evolutionary relationships among Protozoa* (ed. G.H. Coombs, K. Vickerman, M.A. Sleigh and A. Warren), pp. 409-447, Kluwer, Dordrecht.

Curds, C.R. (1963) The flocculation of suspended matter by *Paramecium caudatum*. *J. Gen. Microbiol.* **33**, 357-363.

Curds, C.R. (1969) An illustrated key to the Brithish freshwater ciliated protozoa commonly found in activated sludge. *Wat. Pollut. Res.* Technical paper N°12 I-IV, 1-90.

Curds, C.R. (1975) Protozoa. In *Ecological aspects of used-water treatment* (ed. C. Curds and H.A. Hawkes), vol. 1, Academic Press, London.

Curds, C.R. (1982) British and other freshwater ciliated protozoa. Part I. Ciliophora, Kinetofragminophora. In *Synopsis of the British fauna n° 22* (ed. M. Kermack and R.S.K. Barnes), Cambridge University Press, New York.

Curds, C.R. (1983) British and other freshwater ciliated protozoa. Part II. Ciliophora, Oligohymenophora and Polyhymenophora. In: *Synopsis of the British fauna n° 23.* (ed. M. Kermack and R.S.K. Barnes), Cambridge University Press, New York.

Curds, C.R. (1992) *Protozoa and the Water Industry*, Cambridge University Press, Cambridge.

Curds, C.R. (1993) Interactions protozoaires-bactéries dans les méthodes aérobies de traitement des eaux useés. *Ann. Biol.* **1**, 1-12.

Curds, C.R. and Vandyke, J.M. (1966) The feeding habits and growth rates of some freshwater ciliated found in activated-sludge process. *J. Appl. Ecol.* **3**, 127-138.

Curds, C.R., Coockburn, A. and Vandyke, J.M. (1968) An experimental study of the role of the ciliated protozoa in the activated sluge process. *Water Pollut. Control* **67**, 312-329.

Curds, C.R. and Fey, G.J. (1969) The effect of ciliated protozoa on the fate of *Escherichia coli* in the activated sludge process. *Water Res.* **3**, 853-867.

Curds, C.R. and Cockburn, A. (1970a) Protozoa in biological sewage treatment processes. I. A survey of the protozoan fauna of British percolating filters and activated sludge plants. *Water Res.* **4**, 225-236.

Curds, C.R. and Cockburn, A. (1970b) Protozoa in biological sewage treatment processes. II. Protozoa as indicators in the activated sludge process. *Water Res.* **4,** 237-249.

Curds, C.R. and Wu, I.C.H. (1983) A review of the Euplotidae (Hypotrichida, Ciliophora). *Bull. Br. Mus. Nat. His. Zool.* **31**, 191-247.

Cybis, L.F. de A., and Horan, N.J. (1997) Protozoan and metazoan in Sequencing Batch Reactors (SBR).*Water Sci. Technol.* **35**, 1-81.

Dive, D. (1973) La nutrition holozoique des ciliés. Ses conséquences dans l´épuration naturalle et artificialle. *Ann. Biol.* **1**, 344-380.

Dragesco, J. (1966) Observations sur quelques ciliés libres. *Arch. Protistenkd.* **109**, 155-206.

Dragesco, J. (1970) Ciliés libres du Cameroun. *Annls. Fac. Sci. Univ.féd. Cameroun* (Numéro Hors-Série) **1970**, 1-141.

Duchene Ph. (1991) The most significant protozoa and metazoan currently fond in activated sludge. In *Biological Approach to Sewage: Current Status and Perspectives* (ed. P. Madoni), pp. 33-37, Perugia.

Eikelboom, D.H. and Buijsen, H.J.J. (1981) *Microscopic sludge investigation manual.* TNO Research Institute for Environmental Hygiene, Water and Soil Division, Delft, Netherlands.

Esteban, G., Téllez, C. and Bautista, L.M. (1990) Effects of habitat quality on ciliated Protozoa communities in sewage treatment plants. *Environ. Technol.* **12,** 381-386.

Esteban, G., Téllez, C. and Bautista, L.M. (1991) Dynamics of ciliated protozoa communities in activated sludge process. *Water Res.* **25,** 967-972.

Esteban, G., Téllez, C. and Bautista, L.M. (1992) The indicator value of *Tetrahymena thermophila* populations in the activated-sludge process. *Acta Protozool.* **31**, 129-132.

Esteban, G.S., Clarke, K.J. and Finlay, B.J. (1998) Rapid techniques for the identification of free-living protozoa. In *Techniques in Microbial Ecology*, (ed. R.S. Burlage, R. Atlas, D. Stahl, G. Geesey and G. Sayler), Oxford University Press.

Fenchel, T. (1987) *Ecology of Protozoa. The Biology of Free-Living Phagotrophic Protists*, Springer-Verlag, Berlin.

Fernández-Galiano, D. (1966) Une nouvelle methode pour la mise en evidence de l'infraciliature des cilies. *Protistologica* **2(1)**, 35-37.

Fernández-Galiano, D. (1976) Silver impregnation of ciliated protozoa: procedure yielding good results with the pyridinated silver carbonate method. *T. Am. Microsc. Soc.* **95 (4)**, 557-560.

Fernández-Galiano, D. (1994) The ammoniacal silver carbonate method as a general procedure in the study of protozoa of sewage (and other) waters. *Water Res.* **28**, 495-496.

Fernández-Galiano, D., Guinea, A., Serrano, S., Martín-Cereceda, M., Arregui, L., Rodríguez, B., Campos, I., Calvo, P., and Suarez, J. (1996) *Guía práctica de identificación de protozoos ciliados en estaciones depuradoras de aguas residuales por lodos activos de la Comunidad Autónoma de Madrid.* (ed. Departamento de Microbiología III, Universidad Complutense), Madrid.

Fleury, A., Delgado, P., Iftode, F. and Adoutte, A. (1992) A molecular phylogeny of ciliates: What does it tell us about the evolution of the cytoeskeleton and of developmental strategies?. *Dev. Genet.* **13**, 247-254.

Foissner, W. (1984) Infraciliatur, silberlieniensystem und biometrie einiger neuer und wenig bekannter terrestrischer, limminischer und mariner Ciliaten (Protozoa, Ciliophora) aus den Klassen Kinetofragminophora, Colpodea und Polyhymenophora. *Stapfia* **12**, 1-165.

Foissner, W. (1988) Taxonomic and nomenclatural revision of Sladecek's lists of ciliates (Protozoa, Ciliophora) as indicators of water quality. *Hydrobiologia* **166**, 1-64.

Foissner, W. (1991) Basic light and scanning electron microscopic methods for taxonomic studies of ciliated protozoa. *Eur. J. Protistol.* **27**, 313-330.

Foissner, W. (1992) Evaluating water quality using protozoa and saprobity indexes. In *Protocols in Protozoology*, (ed. Lee, J.J. and Soldo A.T.), Society of Protozoologists.

Foissner, W. and Berger, H. (1996) A user friendly guide to the ciliates (Protozoa, Ciliophora) commonly used by hydrobiologists as bioindicators in rivers, lakes, and wastewaters, with notes on their ecology. *Freshwater Biol.* **35**, 375-482.

Foissner, W., Berger, H., Blatterer, H. and Kohmann, F. (1991) Taxonomische und ökologische Revision der Ciliaten des Saprobiensystems. Band I: Cyrtophorida, Oligotrichida, Hypotrichia, Colpodea. Bayerisches Landesamt für Wasserwirtschaft (Herausgeber und Verlag), München.

Foissner, W., Berger, H. and Kohmann, F. (1992) Taxonomische und ökologische Revision der Ciliaten des Saprobiensystems. Band II: Peritrichia, Heterotrichida, Odontostomatida. Bayerisches Landesamt für Wasserwirtschaft (Herausgeber und Verlag), München.

Foissner, W., Berger, H. and Kohmann, F. (1994) Taxonomische und ökologische Revision der Ciliaten des Saprobiensystems. Band III: Hymenostomata, Prostomatida, Nassulida. Bayerisches Landesamt für Wasserwirtschaft (Herausgeber und Verlag), München.

Foissner, W., Berger, H., Blatterer, H. and Kohmann, F. (1995) Taxonomische und ökologische Revision der Ciliaten des Saprobiensystems. Band IV: Gymnostomatea, Loxodes, Suctoria. Bayerisches Landesamt für Wasserwirtschaft (Herausgeber und Verlag), München.

Fried, J. and Lemmer, H. (2003) On the dynamics and function of ciliates in sequencing batch biofilm reactors. *Water Sci. Technol.* **47**, 189-196.

Güde, H. (1979) Grazing by protozoa as selection factor for activated sludge bacteria. *Microb. Ecol.* **5**, 225-237.

Guhl. Von W. (1979) Beitrag zur systematic, biologie und morphologie der Epistylidae (Ciliata, Peritricha). *Arch. Protistenkd.* **121**, 417-483.

Haussman, K. and Bradbury, P.C. (1996) *Ciliates: cells as organisms*, Gustav Fischer Verlag, Stuttgart.

Haussman, K., Hülsmann, N. and Radek, R. (2003) *Protistology*, 3rd edn., E. Schweizerbart'sche Verlagsbuchhandlung (Nägele u. Overmiller), Stuttgart.

Hoag, G., Widmer, W. and Hovey, W. (1980) Microfauna and RBC performance. Laboratory and full-scale systems. *Procceedings of the First National Symposium on Rotation Biological Contactor Technology*, pp, 167-188.

Honigber, B.M. Balamuth, W., Bovee, E.C., Corliss, J.O., Gojdics, M., Hall, R.P., Kudo, R.R., Levine, N.D., Loeblich, A.R., Weiser, J. and Wenrich, D.H. (1964) A revised classification of the Phylum Protozoa. *J. Protozool.* **11**, 7-20.

Isac, L., Rodriguez, E., Salas, M.D. , Fernández, N., Zornoza, A., Pérez-Uz B., Arregui, L., Calvo, P. and Serrano, S. (2006) Album fotográfico de microorganismos presentes en fangos activos. *Tecnología del Agua.*

Kinner, N.E. and Curds, C.R. (1987) Development of protozoan and metazoan communities in rotating biological contactor biofilms. *Water Res.* **21**, 481-490.

Kinner, N.E, Blackwill, D.L, and Bishop, P.L. (1983) Light and electron microscopic studies of microorganisms growing in rotating biological contactor biofilms. *Appl. Environ. Microb.* **45**, 1659-1669.

Kinner, N.E., Curds, C.R. and Meeker, L.D. (1988) Protozoa and metazoan as indicators of effluent quality in rotating biological contactors. *Water Sci. Technol.* **20**, 199-204.

Kirby, H. (1945) The structure of the common intestinal trichomonad of man. *J. Parasitol.* **31**, 163-174.

Klein, B.M. (1926) Über eine neue Eigentümlichkeit Pellicula von *Chilodon uncinatus* Ehrbg. *Zool. Anz.* **67**, 1-2.

Klein, B.M. (1958) The "dry" silver method and its proper use. *J. Protozool.* **5**, 99-103.

Lee, J.J. and Soldo A.T. (1992) *Protocols in Protozoology*, (ed. Society of Protozoologists), Lawrence, Kansas.

Lee, J.J., Leedale, G.F. and Bradbury, P. (2000) An *illustrated guide to the Protozoa*, 2nd edn, (ed. Society of Protozoologists) vol. I and II, Lawrence, Kansas.

Lee, S., Basu, S., Tyler, C.W. and Wei, I.W (2004) Ciliate populations as bio-indicators at Deer Island treatment plant. *Adv. Environ. Res.* **8**, 371-378.

Levine, N.D., Corliss, J.O., Cox, F.E.G., Deroux, G., Grain, J., Honigberg, B.M., Leedale, G.F., Loeblich III, A.R., Lom, J., Lynn, D., Merinfeld, E.G., Page, F.C., Poljanski, G., Sprague, V., Vavra, J. and Wallace, F.G. (1980) A newly revised classification of the Protozoa. *J. Protozool.* **27**, 37-58.

Lipscomb, D.L., Farris, J.S., Kallersjo, M. and Tehler, A. (1998) Support, ribosomal sequences and the phylogeny of the eukaryotes. *Cladistics* **14**, 303-338.

Luna-Pabello, V.M., Mayen, R., Olvera-Viascan, V., Saavedra, J. and Duran de Bazua, C. (1990) Ciliated protozoa as indicators of a wastewater treatment system performance. *Biol. Waste* **32**, 81-90.

Luna-Pabello, V.M., Aladro-Lubel, M.A. and Durán de Bazúa, C. (1996) Biomonitoring of wastewaters in treatment plants using ciliates. *J. Ind. Microbiol.* **17**, 62-68.

Lynn, D.H. (1996) My journey in ciliates systematics. *J. Eukaryot. Microbiol.* **43**, 253-260.

Lynn, D.H. and Corliss, J.O. (1991) Ciliophora. In *Microscopic anatomy of invertebrates*, (ed. W. Harrison), pp. 333-467, Wiley-liss, New York.

Lynn, D.H. and Small, E.B. (2000) Phylum Ciliophora. In*: An illustrated guide to the Protozoa*, 2nd edn (ed. J.J. Lee, G.F. Leedale and P. Bradbury), vol. I, Society of Protozoologists, Lawrence, Kansas.

Ma, H.W., Choi, J.K. and Song, W.B. (2003) An improved silver carbonate impregnation for marine ciliated protozoa. *Acta Protozool.* **42 (2)**, 161-164.

Madoni, P. (1982) Growth and succession of ciliate populations during the establishment of a mature activated sludge. *Acta Hydrobiol.* **24**, 223-232.

Madoni, P. (1984) Estimation of the size of freshwater ciliate populations by a sub-sampling technique. *Hydrobiologia* **111**, 201-206.

Madoni, P. (1988) *I protozoi ciliate nel controllo di efficienza dei fanghi attivi.* Centro Italiano Studi di Biologia Ambientale, Reggio Emilia.

Madoni, P. (1991) Role of Protozoans and their indicator value in the activated sludge process. In *Biological Approach to Sewage: Current Status and Perspectives* (ed. P. Madoni), pp. 21-27, Perugia.

Madoni, P. (1994a) A sludge biotic index (SBI) for the evaluation of the biological performance of activated sludge plants based on the microfauna analysis. *Water Res.* **28**, 67–75.

Madoni, P. (1994b) Microfauna biomass in activated-sludge and biofilm. *Water Sci. Technol.* **29**, 63-66.

Madoni, P. (1994c) *Manuale di applicazione. La microfauna nell'analisi di qualitá biologica dei fanghi attivi. Indice biotico del fango: SBI.* Azienda Gas Aqua Consortiale di Reggio Emilia. Universitá degli Studi di Parma.

Madoni, P. (2000) The acute toxicity of nickel to freshwater ciliates. *Environ. Pollut.* **109**: 53-59.

Madoni, P. and Ghetti, P.F. (1981) The structure of ciliated protozoa communities in biological sewage-treatment plants. *Hydrobiologia* **83**, 207-215.

Madoni, P. and Antonietti, R. (1984) Colonization dynamics of ciliated protozoa populations in an activated sludge plant. In *4° Symposio Dinamica Popolazioni*, pp. 105-112, Parma.

Madoni, P. and Romeo, M.G. (2006) Acute toxicity of heavy metals towards freshwater ciliated protist. *Environ. Pollut.* **141**, 1-7.

Madoni, P., Ghetti, P.F. and Antonietti, R. (1979) Analisi delle caratteristiche biologiche di un impianto piloto a tamburo rotante. Dinamica di colonizzazione di ciliati e altri gruppi. *Inquinamento* **4**, 1-14.

Madoni, P., Davoli, D. and Chierici, E. (1993) Comparative analysis of the activated sludge microfauna in several sewage treatment works. *Water Res.* **27**, 1485-1491.

Madoni, P., Davoli, D., Gorbi, G. and Vescovi, L. (1996) Toxic effect of heavy metals on the activated sludge protozoan community. *Water Res.* **30** (1), 135-141.

Margulis L., Corliss J.O., Melkonian M. and Chapman, D.J. (1990) *Handbook of Protoctista : the structure, cultivation, habitats and life histories of the eukaryotic microorganisms and their descendants exclusive of animals, plants and fungi: a guide to the algae, ciliates, foraminifera, sporozoa, water molds, slime, molds and the other protoctists*, Jones and Barlett Pub, Boston.

Martín-Cereceda, M., Serrano, S. and Guinea, A. (1995) Occurrence of *Acineria uncinata* in activated sludge plants: Morphology and relationship with physico-chemical parameters. *Arch. Protistendk.* **146**: 79-84.

Martín-Cereceda, M., Serrano, S. and Guinea, A. (1996) A comparative study of ciliated protozoa communities in activated-sludge plants. *FEMS Microbiol. Ecol.* **21**, 267-276.

Martín-Cereceda, M., Serrano, S. and Guinea, A. (2001a) Biofilm communities and operational monitoring of a rotating biological contactor system. *Water Air Soil Poll.* **126**, 193-206.

Martín-Cereceda, M., Perez-Uz, B., Serrano, S., Guinea, A. (2001b) Dynamics of protozoan and metazoan communities in a full scale wastewater treatment plant by rotating biological contactors. *Microbiol. Res.* **156**, 225-238.

Martín-Cereceda, M., Alvarez-Barrientos, A., Serrano, S. and Guinea, A. (2001c) Confocal and light microscope examination of protozoan and other microorganisms in the biofilms from a rotating biological contactor wastewater treatment plant. *Acta Protozool.* **40**, 263-272.

Martín-Cereceda, M., Pérez-Uz, B., Serrano, S. and Guinea, A. (2001d) An integrated approach to analyse biofilms of a full scale wastewater treatment plant. *Water Sci. Technol.* **46**, 199-206.

Martín-Cereceda, M., Zamora, J., Pérez-Uz, B. and Guinea, A. (2002) Ciliate communities of rotating biological contactor biofilms: A multivariate approach. *Syst. Appl. Microbiol.* **25**, 301-313.

Mas-Acebes, M. (2007) La cadena trófica en los sistemas de depuración biológicos. ¿Quién se come a quién?. *Tecnología del Agua* **285**, 30-40.

Montagnes, D.J.S and Lynn, D.H. (1987) A quantitative protargol stain (QPS) for ciliates: method description and test of its quantitative nature. *Mar. Microb. Food Webs* **2(2)**, 83-93.

Moskowitz, N. (1950) The use of protein silver for staining protozoa. *Stain Technol.* **25**, 17-20.

Nicolau, A., Dias, N., Mota, M. and Lima, N. (2001) Trends in the use of protozoa in the assessment of wastewater treatment. *Res. Microbiol.* **152**, 621-630.

Olmo, J.L. and Téllez, C. (1997) The ammoniacal silver carbonate method applied to oxytrich ciliates. *Quekett J. Microsc.* **38**, 3-5.

Pérez-Uz, B., Franco, C., Martín-Cereceda, M., Arregui, L., Campos, I., Serrano, S., Guinea, A. and Fernández-Galiano, D. (1998) Biofilm characterization of several wastewater treatment plants with rotating biological contactors in Madrid (Spain). *Water Sci. Technol.* **37**, 215-218.

Poole, J.E.P. (1984) A study of the relationship between the mixed liquor fauna and plant performance for a variety of activated sludge sewage treatment works. *Water Res.* **18**, 281-287.

Pretorius, W.A. (1971) Some operational characteristics of a bacterial disc unit. *Water Res.* **5**, 1141-1146.

Puigagut, J., Salvadó, H., García, D., Granesc, F. and García, J. (2007) Comparison of microfauna communities in full scale subsurface flow constructed wetlands used as secondary and tertiary treatment. *Water Res.* **41**, 1645-1652.

Puytorac, P.de, Grain, J. and Mignot, J.P. (1987) *Précis de Protistologie*, Société Nouvelle des Éditions Boubée, Paris.

Ratsak, C.H., Hooi, B.W. and Van Verseveld, H.W. (1994) Biomass reduction and mineralization increase due to the ciliate *Tetrahymena pyriformis* grazing on the bacterium *Pseudomonas fluorescens*. *Water Sci. Technol.* **29**, 119-128.

Roberts, D.M. and Causton, H. (1988) Silver nitrate impregnation of ciliated protozoa. *Arch. Protistenk.* **135**, 299-318.

Salvadó, H (1994) Effect of mean cellular retention time on ciliated protozoa populations in urban wastewater treatment plants based on a proposed model. *Water Res.* **28(6)**, 1315-1321.

Salvadó, H. (2001) La microscopía como técnica para optimizar el control del tratamiento biológico de las aguas residuales mediante fangos activos. In *Ciencia y Tecnología*, vol. 2, pp. 177-185, Ediciones Tibidabo, Barcelona.

Salvadó, H., Gracia, M.P. and Amigó, J.M. (1995) Capability of ciliated Protozoa as indicators of effluent quality in activated sludge plants. *Water Res.* **29**, 1041-1050.

Schlegel, M. (1994) Molecular phylogeny of eukaryotic. *Trends. Ecol. Evol.* **9**, 330-335.

Selivanovskaya, S.Y., Petrov, A.M., Egorova K.V. and Naumova R.P. (1997) Protozoan and metazoan communities treating a simulated petrochemical industry wastewater in a rotating disc biological reactor. *World J. Microb. Biot.* **13**, 511-517

Sleigh, M.A. (1989) *Protozoa and other Protists* (ed. E. Arnold), London.

Small, E.B. (1973) A study of ciliate protozoa from a small polluted stream in East- Central Illinois. *Am. Zool.* **13,** 225-230.

Sogin, M.L. (1991) Early evolution and the origin of eukaryotes. *Curr. Opin. Genet. Dev.* **1**, 457-463.

Sommer, U. (1996) Plankton Ecology: The past two decades of progress. *Naturwissenschaften* **83**, 293-301.

Torpey W.N., Heukelekian H., Kaplowsky A.J. and Epstein R. (1971) Rotating disks with biologicas growths prepared wastewater for disposal and reuse. *J. Water Pollut. Con. F.* **43**, 2181-2188.

Vymazal, J., Sladecek, V. and Stach, J. (2001) Biota participating in wastewater treatment in a horizontal flow constructed wetland. *Water Sci. Technol.* **44(11-12)**, 211-214.

Warren, A. (1986) A revision of the genus *Vorticella* (Ciliophora, Peritrichida). *Bull. Br. Mus. Nat. His. Zool.* **50**, 1-57.

Watson, J.M. (1945) Mechanisms of bacterial flocculation caused by protozoa. *Nature* **155**, 271.

Wicklow B.J. (1997) Signal-induced defensive phenotypic changes in ciliated protists: Morphological and ecological implications for predator and prey. *J. Eukaryot. Microbiol.* **44(3)**, 176-188.

Wilbert, N. (1975) Eine verbesserte Technik der Protargol-imprägnation für Ciliaten. *Mikrokosmos*, **6**, 171-179.

Wu, I.C.H. and Curds, C. (1979) A guide to the species of the genus *Aspidisca*. *Bull. Br. Mus. Nat. His. Zool.* **28**, 1-34.

Web Sites

http://www.nies.go.jp/chiiki1/protoz/
http://starcentral.mbl.ed/microscope
http://protist.i.hosei.ac.jp/PDB/Images/Protista.html

Index

Page numbers in *italics* represent figures or tables. Where a page contains both figure and text, the page number is given both in plain type and italics.

by Susana Serrano, Lucia Arregui, Blanca Perez-Uz, Pilar Calvo and Almudena Guinea.
ISBN: 9781843391715. Published by IWA Publishing, London, UK.

IWA Publishing's authorised EU representative for General Product Safety Regulations is Diane D'Arras, 15 rue Duret, 75116 Paris, France, e-mail: safety@iwap.co.uk.

Printed and bound by CPI Group (UK) Ltd, Croydon, CR0 4YY
17/09/2026
02218812-0001